NextGen
Genealogy

NextGen Genealogy

The DNA Connection

David R. Dowell

AN IMPRINT OF ABC-CLIO, LLC

Santa Barbara, California • Denver, Colorado • Oxford, England

Library of Congress Cataloging-in-Publication Data

Dowell, David R.
  Nextgen genealogy : the DNA connection / David R. Dowell.
    pages cm
  Includes bibliographical references and index.
    ISBN 978–1–61069–727–9 (pbk. : alk. paper) — ISBN 978–1–61069–728–6 (ebook) 1. Genetic genealogy. 2. DNA—Analysis. I. Title.
CS21.D89   2015
929.1072—dc23   2014026726

ISBN: 978–1–61069–727–9
EISBN: 978–1–61069–728–6

19 18 17 16 15   1 2 3 4 5

This book is also available on the World Wide Web as an eBook.
Visit www.abc-clio.com for details.

Libraries Unlimited
An Imprint of ABC-CLIO, LLC

ABC-CLIO, LLC
130 Cremona Drive, P.O. Box 1911
Santa Barbara, California 93116-1911

This book is printed on acid-free paper ∞

Manufactured in the United States of America

Contents

Preface

NextGen Genealogy: The DNA Connection is intended to help you appreciate the four separate patterns by which men inherit the four discrete groupings of their DNA. The use of the word "men" in the previous sentence was deliberate: women inherit through only three of these processes. Being able to differentiate between these patterns of how DNA is passed from parents to child is essential to understanding which test(s) to take for genealogical purposes and how to interpret the results. Used properly, they help us find connections with direct-line ancestors and the cousins who are their descendants.

This book does not come with a money-back guarantee. The family history research of most readers will benefit if you learn to combine the information from within your cells with the information you collect from traditional research. Some of you will be disappointed—at least at first. A poor outcome is likely to result from one or more of the following issues:

1. *Your pedigree chart is not robust enough.* Build it. If you are waiting for DNA test results, use the interval to apply documentary research and to extend all of your direct lines back as far as possible but at least eight generations. You may never quite complete this task, but it is a realistic goal. (After decades of research, my own tree still breaks down on a couple of lines after only five generations.) Continue to build your tree as you analyze your DNA test results. You cannot understand where to fit DNA matches into your family unless you have the context provided by a reasonably well-developed family tree.
2. *Extended family/cousins have not been tested.* Recruit them. Lots of people have taken DNA tests, but they still make up a miniscule percentage of the earth's inhabitants. Test takers are not evenly distributed throughout all groups. Be proactive in expanding DNA databases to include those who have a high probability of matching you.
3. *You do not know how to fish the information out.* Read on. This book will not provide you with all the techniques you will ever need to know, but it should give you enough to get started.

Chapter 1 will provide you with most of the basic genetic concepts and terms you will need to start practicing genetic genealogy. If you do not retain all of its content on first reading, come back later when you have a specific need for a review.

Chapters 2 through 5 will introduce you to yDNA, mtDNA, atDNA, and xDNA—the four types of human DNA that can be useful to us as genetic genealogists. Each of these chapters will help you understand the unique inheritance pattern of one of these types of DNA and appreciate how you can begin to apply test results in your own family research.

Chapter 6 extends your voyage of family history discovery into even earlier eras of your deep ancestry.

Chapter 7 raises questions about whether we should test our DNA and how each of us may arrive at different conclusions when ethical issues arise.

Chapter 8 discusses what is coming next and gives suggestions for additional learning experiences as you continue your journey into this fast-evolving field.

At the end of this book, you will find recommended reading for further learning, a glossary of terms you may encounter, and a comprehensive index to help you single out specific concepts or terms of interest.

For genealogists, the actual results listed on DNA test reports have little importance. The meaning comes as you make comparisons with others who share (or do not share) part of the information recorded inside your cells. Mastering techniques that allow you to make these assessments will enable you to benefit from adding DNA testing to your genealogical tool kit.

Some readers will be the first of your family to take some of the tests at some of the companies. As a consequence, you may be disappointed with your lack of matches. While this may be the case with results from a Y-chromosome or mitochondrial test, I have yet to meet individuals who did not have matches on their autosomal DNA. Finding the individuals with whom you share DNA and, therefore, a relationship is the easy part. The hard work comes in finding where in your family tree your shared ancestor is perched. This book offers suggestions about how you can choose the test that will give you the best chance of tearing down long-standing genealogical brick walls. It will also help you get started down the sometimes long and tortuous path of making sense of your results and interpreting the matches you get with others.

THE BIRTH OF THIS BOOK

This book was originally the offspring of a genetic match. A few years ago, I discovered that I shared a mitochondrial line with genetic genealogist CeCe Moore. We may never discover who our common female ancestress is, but you will hear more about this match later in the book. After we got acquainted, we decided to write a book to introduce genealogists to the 21st-century tool of genetic genealogy. Subsequently, life and CeCe's success as a DNA advisor to television productions intervened and she had to step aside from this project.

At a late stage of the project, Angie Bush came to my rescue and helped me get this manuscript over the finish line. Angie poked holes in my thought process throughout the book. She also helped me identify and fill gaps. She is responsible for strengthening the book, but I alone am responsible for the content and any shortcomings it may have.

While Angie as a molecular genealogist has an extensive background in both genetics and genealogy, I have never taken an academic course in the life sciences. God has decided to punish me for that deficiency by inflicting me with a vast curiosity about how we can and should apply the information discovered inside our bodies by genetic testing to family history research and personal health decisions.

Although this book was written for those of you who have little or no background in genetics, it is assumed that you have a basic knowledge of traditional genealogical principles. If that is not the case, you may want to do some reading in that area, such as "Backward Thinking and Other Keys to Successful Genealogical Research" (Chapter 2 in my *Crash Course in Genealogy*).[1]

The primary audience for this book is not the academic community. I have deliberately chosen to communicate in the same breezy style employed in blogs. I have decided to ignore the advice of Mary Jane Frances Smith: "The style used to write on a blog, in an e-mail, or in other forums on the Internet is not the style a writer should use in a letter of reference, print magazine, professional journal, or book."[2] If that decision bothers you, you are probably not the intended audience whom I had in mind for this book. There is a time and place for every style. This is a book intended for novices in this exciting new field. It is not a book intended to advance the frontiers of discovery for seasoned experts.

ACKNOWLEDGMENTS

Although writing often seems to be a solitary endeavor, I could not be successful at it without both professional and personal support teams. I wish acknowledge some of them here.

Jon Dowell served as the legal department for dealing with contract issues, as he has done in all my publishing endeavors.

CeCe Moore offered critical advice on early drafts of some chapters.

Nir Leibovich and Bennett Greenspan from Gene-by-Gene gave permission to share their haplogroup migration maps.

Richard Spinello influenced my thinking on cyberethics and graciously allowed me to use his diagram of "interested" parties in cyberspace issues.

Blanche Woolls, Barbara Ittner, and Emma Bailey provided their usual wise counsel and encouragement in navigating the publishing process.

Denise Dowell proofread the entire manuscript and only rarely interrupted my early-morning "quiet times" when I can best concentrate on writing.

Elise Friedman and Jim Turner shared tools to help me better understand X-chromosome DNA.

RaggZ showed up at least once during almost every writing session to lower my blood pressure by a combination of purrrrfect utterances and multiple shin rubs—although there also were occasional catty comments when treats were slow to arrive.

NOTES

1. David R. Dowell, *Crash Course in Genealogy*. Santa Barbara, CA: Libraries Unlimited, 2011, pp. 13–33.

2. Mary Jane Frances Smith, "Review of *Basics of Genealogy Reference: A Librarian's Guide* by Jack Simpson," *Association of Professional Genealogists Quarterly* (March 2010): 14.

1

What Is DNA? Family Information inside Our Cells

When did you last say "*WOW*"—or even like a big kid say, "Yippee!" I have said this many times in the last 12 months since becoming involved in DNA testing to find answers to fill those missing gaps in my genealogy. DNA results and your paper trail go together in filling the gaps in your family tree!

—Jan Kelly, enthusiastic genetic genealogist[1]

The body has a long memory indeed. Written in the quirky tongue of DNA and wound into the nucleus of nearly every human cell are biological mementos of the family who came before us. And science is finding ways to dig them out, rummaging through our DNA as if it were a trunk in the attic.

—Journalist Carolyn Abraham[2]

That "quirky tongue" uses only four letters—A, C, G, and T—to record our genomic code across 3 billion base pairs that are replicated in each cell in our bodies except for our red blood cells.

Don't allow yourself to be overwhelmed by the brief introduction to genetics offered in this chapter. Ultimately, all the knowledge that you need to retain to succeed as a novice genetic genealogist is the ability to differentiate between the four distinct inheritance patterns of the four types of DNA and the ability to develop and execute a plan to evaluate matches you may encounter with each of these types. You will be reminded of these four types of inheritance at the end of this chapter. Even experienced genetic genealogists often have to remind themselves of these differences from time to time. Understanding these differences will be the key to your success as you learn to apply the results from DNA testing to your family history research. As molecular genealogist Angie Bush reminds us, "The focus should be on what the different types of DNA are and how they are passed from parents to children. Understanding those principles is what allows DNA to be used as a genealogical record to solve questions of kinship and identity."[3]

Tests on four separate and distinct parts of our human DNA provide discrete but complementary information to help us fill in gaps in our family histories. Y-chromosome DNA (yDNA) and mitochondrial DNA (mtDNA) are relatively stable and change very slowly through many intergenerational transfers. They can help us fill in gaps in our knowledge of our deep ancestry, but each can inform us about only one direct line on our pedigree charts. In contrast, autosomal DNA (atDNA) changes very significantly and somewhat unpredictably from any one generation to the next. It can help us sort out relationships on all lines of our pedigree charts, but its reach is generally not as deep. The complexity of X-chromosome inheritance (xDNA) makes it the most difficult type of DNA to utilize for genealogical purposes, but it can provide valuable insights in many circumstances. Taken together, these tests provide knowledgeable genealogists with a choice of tools to use to fill specific gaps in their family histories. Interpreting the results of tests of each of these four kinds of DNA will be discussed in detail in succeeding chapters. However, first let us turn our attention to a quick discussion of just what is DNA.

DNA has immense power to help us understand who we are, where our descendants are headed, and from whence our ancestors came. It is on the last of these three potentials that this book will focus. DNA testing is an important new tool for 21st-century genealogists. It has been said that DNA testing is "putting the gene back in genealogy." It is a supplement to—but not a replacement for—traditional family history research. Wrapping your mind around some of the information in this book may take some time and probably will require some rereading. However, you are not being asked to become a geneticist. Rather, you will be given information on how to integrate DNA test results with what you already know through the traditional process of family history research.

Discussions about DNA testing often have a conjectural elephant in the room with characteristics as diverse as did the physical beast visited by the six blind men of mythical Indostan (Figure 1.1):

Figure 1.1
Your perspective determines what you perceive. Is there an elephant in the room? Source: Illustrator unknown, "Blind Men and an Elephant." From Charles Maurice Stebbins and Mary H. Coolidge, *Golden Treasury Readers: Primer* **(New York, NY: American Book Company, 1909),** *89*

The Blind Men and the Elephant
John Godfrey Saxe (1816–1887)

It was six men of Indostan
To learning much inclined,
Who went to see the Elephant
(Though all of them were blind),
That each by observation
Might satisfy his mind.

The First approached the Elephant,
And happening to fall
Against his broad and sturdy side,
At once began to bawl:
"God bless me! but the Elephant
Is very like a WALL!"

The Second, feeling of the tusk,
Cried, "Ho, what have we here,
So very round and smooth and sharp?
To me 'tis mighty clear
This wonder of an Elephant
Is very like a SPEAR!"

The Third approached the animal,
And happening to take
The squirming trunk within his hands,
Thus boldly up and spake:
"I see," quoth he, "the Elephant
Is very like a SNAKE!"

The Fourth reached out an eager hand,
And felt about the knee
"What most this wondrous beast is like
Is mighty plain," quoth he:
"'Tis clear enough the Elephant
Is very like a TREE!"

The Fifth, who chanced to touch the ear,
Said: "E'en the blindest man
Can tell what this resembles most;
Deny the fact who can,
This marvel of an Elephant
Is very like a FAN!"

The Sixth no sooner had begun
About the beast to grope,
Than seizing on the swinging tail
That fell within his scope,
"I see," quoth he, "the Elephant
Is very like a ROPE!"

And so these men of Indostan
Disputed loud and long,
Each in his own opinion
Exceeding stiff and strong,
Though each was partly in the right,
And all were in the wrong![4]

So what is this elephant in the room we call DNA?

- To a chemist, it is deoxyribonucleic acid.
- To a geneticist, it is a code that differentiates different kinds of organisms.
- To our bodies, it is an infinitely detailed instruction book on how to create our specific physical body, change it over time, and if necessary to repair it.
- To medical researchers, it provides aggregates of anonymous samples that can be studied to discover associations between a specific gene or groups of genes and certain health conditions.
- To physicians, it can predict how individual humans will respond/react to certain drugs and is a guide to a new era of personalized genomic medicine, in which it is not assumed that one solution fits all patients even if they suffer from the same malady.
- To someone wanting to be entertaining at a cocktail party, it offers an ever-changing source of trivia.
- To a criminal investigator, it helps identify, confirm, or eliminate suspects.
- To an anthropologist, it helps trace the migration of humans through scores of millennia.
- To a genealogist, it is a tool to confirm or refute information we have from oral and documentary sources and to guide us by pointing to previously unknown relationships.
- To an ethicist, its use raises all kinds of new questions—many of which we long will be struggling to answer.

The purpose of this book is not to make you knowledgeable about genetics, except as far as you need to know about the possibilities and limitations in applying DNA testing to further family history research. Just as you do not have to be an aeronautical engineer to use an airplane to visit your family, so you do not need to be a geneticist to use the results of DNA testing to extend your family history research.

That being said, a rudimentary understanding of the basics of genetics will allow you to maximize the genealogically relevant information you can extract from DNA test results. Since your author is not a geneticist, the next section relies heavily on the *Genetics Home Reference: Your Guide to Understanding Conditions*[5] to ensure definitions do not unintentionally misstate accurate science. All of the following basic descriptions in this chapter are stripped down to the minimum of what you will need to understand to apply the results of your DNA tests to advance your family history efforts. This free online reference source is provided by the U.S. National Library of Medicine (NLM), which is part of the National Institutes of Health (NIH; http://ghr.nlm.nih.gov). The following basic introduction will introduce basic vocabulary and concepts. Readers wishing more detailed information, particularly health-related information, are advised to visit this robust resource and the links it provides to authoritative external resources.

WHAT IS A GENOME?

A genome is an organism's complete set of DNA, including all of its genes. Each genome contains all of the information needed to build and maintain that organism.

Figure 1.2
Typical human cell showing locations of DNA in the nucleus and the mitochondria (Reprinted with permission from Centre for Genetics Education.)

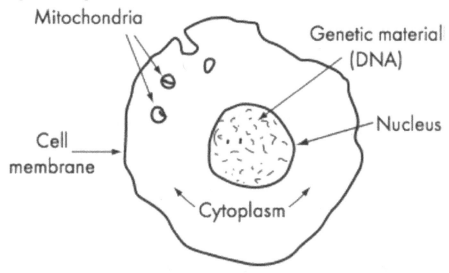

In humans, a copy of the entire genome—more than 3 billion DNA base pairs—is contained in all cells that have a nucleus.[6]

DNA IN ALMOST EVERY HUMAN CELL

DNA can be found in almost every human cell. Figure 1.2 shows a simplified version of a typical cell.[7] It shows the location of the nucleus and the mitochondria, which are the sites where our DNA can be found.

WHAT IS DNA?

DNA, or deoxyribonucleic acid, is the hereditary material in humans and almost all other organisms. Nearly every cell in a person's body has the same DNA. Most DNA is located in the cell nucleus (where it is called nuclear DNA), but a small amount of DNA can also be found in the mitochondria (where it is called mitochondrial DNA or mtDNA).

An important property of DNA is that it can replicate, or make copies of itself. Each strand of DNA in the double helix can serve as a pattern for duplicating the sequence of bases. This is critical when cells divide because each new cell needs to have an exact copy of the DNA present in the old cell.[8]

The familiar double helix is shown in Figure 1.3, along with the four chemical bases that make up the components of DNA.

WHAT IS MITOCHONDRIAL DNA?

Although most DNA is packaged in chromosomes within the nucleus, mitochondria also have a small amount of their own DNA. This genetic material is known as mitochondrial DNA or mtDNA.

Figure 1.3
Segment of the double helix and the four chemical bases that make up DNA

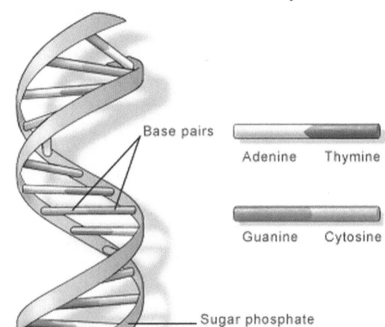

U.S. National Library of Medicine

Mitochondria are structures within cells that convert the energy from food into a form that cells can use. Each cell contains hundreds to thousands of mitochondria, which are located in the fluid that surrounds the nucleus (the cytoplasm).[9]

Although the mitochondria shown in Figure 1.4[10] are very small, they are likely to survive longer than the DNA inside the nucleus of a cell because almost all of our approximately 10 trillion cells contain hundreds and sometimes thousands of copies. Therefore, mitochondrial DNA often can be used to identify human remains long after nuclear DNA can no longer be recovered.

WHAT ARE CHROMOSOMES?

In the nucleus of each cell, the DNA molecule is packaged into thread-like structures called chromosomes. Each chromosome is made up of DNA tightly coiled many times around proteins called histones that support its structure.

Figure 1.4
Cross section of a cell showing the nucleus and surrounding mitochondria

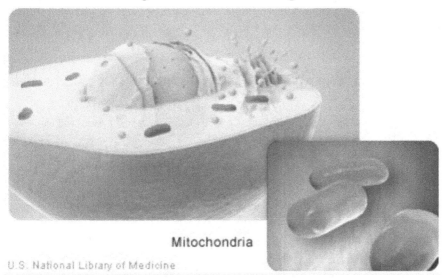

Mitochondria

U.S. National Library of Medicine

Chromosomes are not visible in the cell's nucleus, not even under a microscope, when the cell is not dividing. However, the DNA that makes up chromosomes becomes more tightly packed during cell division and is then visible under a microscope. Most of what researchers know about chromosomes was learned by observing chromosomes during cell division.[11]

Figure 1.5
Relative sizes of the 46 chromosomes

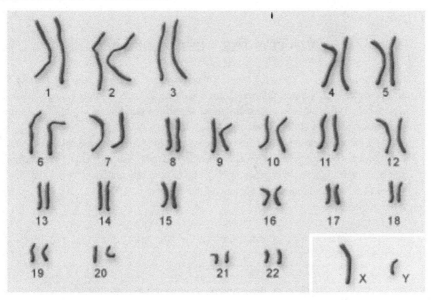

autosomes   sex chromosomes

U.S. National Library of Medicine

The illustration from the National Library of Medicine (Figure 1.5) shows the relative size of the 46 chromosomes. They are organized into 22 pairs of autosomal chromosomes and one pair of sex chromosomes. The latter pair consists of two X chromosomes for women or one X chromosome and one Y chromosome for men. Figure 1.5 depicts the chromosomes for men.[12]

WHAT ARE THE AUTOSOMES AND THE SEX CHROMOSOMES?

In humans, each cell normally contains 23 pairs of chromosomes, for a total of 46. Twenty-two of these pairs, called autosomes, look the same in both males and females. The 23rd pair, the sex chromosomes, differs between males and females.[13]

WHAT IS THE X CHROMOSOME?

The X chromosome is one of the two sex chromosomes in humans (the other is the Y chromosome). The sex chromosomes form one of the 23 pairs of human chromosomes in each cell. The X chromosome spans about 155 million DNA building blocks (base pairs) and represents approximately 5 percent of the total DNA in cells.

Each person normally has one pair of sex chromosomes in each cell. Females have two X chromosomes, while males have one X and one Y chromosome.

Identifying genes on each chromosome is an active area of genetic research. Because researchers use different approaches to predict the number of genes on each chromosome, the estimated number of genes varies. The X chromosome likely contains between 900 and 1,400 genes. These genes perform a variety of different roles in the body.

Genes on the X chromosome are among the estimated 20,000 to 25,000 total genes in the human genome.[14]

Throughout the balance of this book, DNA along the X chromosome will be referred to as xDNA.

WHAT IS THE Y CHROMOSOME?

The Y chromosome is one of the two sex chromosomes in humans (the other is the X chromosome). The sex chromosomes form one of the 23 pairs of human chromosomes in each cell. The Y chromosome spans about 58 million building blocks of DNA (base pairs) and represents almost 2 percent of the total DNA in cells.

The Y chromosome likely contains between 70 and 200 genes. Because only males have the Y chromosome, the genes on this chromosome tend to be involved in male sex determination and development. Other genes on the Y chromosome are important for male fertility.

Genes on the Y chromosome are among the estimated 20,000 to 25,000 total genes in the human genome.[15]

Since the Y chromosome is found only in the genome of males, it may be useful to think of it as the "guY chromosome."

WHAT IS A GENE?

A gene is the basic physical and functional unit of heredity. Genes, which are made up of DNA, act as instructions to make molecules called proteins. In humans, genes vary

in size from a few hundred DNA bases to more than 2 million bases. The Human Genome Project has
estimated that humans have between 20,000 and 25,000 genes.

Every person has two copies of each gene, one inherited from each parent. Most genes are the same in all people, but a small number of genes (less than 1 percent of the total) are slightly different between people. Alleles are forms of the same gene with small differences in their sequence of DNA bases. These small differences contribute to each person's unique physical features.[16]

Most genes are located in the chromosomes along our double helix (see Figure 1.6)[17]. Each chromosome contains many genes; the mitochondria contain a few as well.

Genetics is a relatively new branch of science, and DNA exploration is one of its newer subfields. It is currently one of the most dynamic areas of scientific research, with new discoveries seemingly announced every week. Scientist Bryan Sykes, author of *Seven Daughters of Eve*, helps set the stage: "Although scientists have been using genetics to interpret the past for almost a hundred years, it is only in the last two decades that DNA itself has been recruited to the quest."[18] The first draft of the map of the entire human genome was completed only about a decade ago. This monumental accomplishment resulted from an immense public and private competition known as the Human Genome Project (HGP).

Like early maps of the world, the first draft of the human genome was a bit vague in many of the details. To visualize this project, think of the HGP as being similar to Google's first attempt to map all the structures on the earth from satellite images. The result gave us an idea of where various structures were located and how many there were. However, it didn't tell us very much about the structures themselves, including how they were constructed and which functions they perform. Geneticists continue to fill in those details at a rapid pace by identifying the structures and functions of our cells. New enhancements of that draft seem to be announced daily. What we think we know today is likely to change within a few weeks as more is discovered in the labs.

Figure 1.6

Each chromosome contains many genes

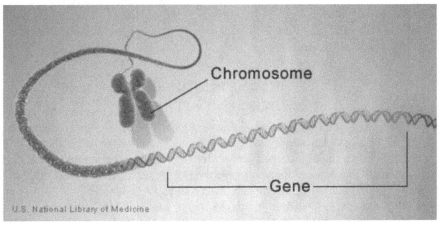

U.S. National Library of Medicine

SHOULD ETHICAL CONSTRAINTS BE IMPOSED?

The HGP was much more than a huge leap in our understanding of human genetics. Similar to the phenomenon of the Space Race in the middle of the 20th century, the HGP accelerated the development of technology to support the mapping process. As ever more of this process is automated, more segments of more human genomes can be sequenced with increasing speed and at lower costs.

Simultaneously, our capacity to store and access bits of information is growing exponentially. Both the size of our databases and the efficiency of our software at correlating those data expand almost at warp speed. This allows us to store information on each human and to correlate it with similar information on each other human in ways that a decade or two ago were only in the imagination of a few science fiction authors.

This combined expansion of genetic knowledge, sequencing ability, and computer capacity is rapidly bringing down the cost of DNA analysis even as we expand the focus of our areas of investigation. We can examine more human DNA now in less time and at a lower cost. The end or even the slowing of these growth curves is not in sight.

The phenomenal growth curves discussed in the previous paragraph are critical to our success as genetic genealogists. Sequencing DNA is still no trivial undertaking. Most of you are aware of Moore's law, which was based on the observation back in 1965 of Intel co-founder Gordon Moore that computer capacity doubled every two years. Even this impressive growth curve has trouble keeping up with the needs of those building genetic databases. Computer scientist Ben Langmead calls this a "DNA data deluge." He reports, "From 2008 to 2013, sequencer throughput increased three to five times per single year."[19]

Langmead visualizes the process:

> A DNA sequencer doesn't produce a complete genome that researchers can read like a book, nor does it highlight the most important stretches of the vast sequence. Instead, it generates something like an enormous stack of shredded newspapers, without any organization of the fragments. The stack is far too large to deal with manually, so the problem of sifting through all the fragments is delegated to computer programs. A sequencer, like a computer, is useless without software.[20]

As with any expanding technology, we must ask the question, should we do it just because we can? Technology expands much faster than our ability to control its consequences or even to predict what they might be. Are there limits or guidelines that should be imposed? Which human values are being impacted? These and similar questions will be revisited later in this book.

IS SOME DNA JUNK?

Initially it was thought that only the genes were important in human inheritance, in proper functioning, and if necessary in regeneration. The rest of the genome was considered to be junk DNA. It was thought that this junk DNA was the debris left over from millennia of human evolution and was no longer vital to creating and maintaining healthy humans. Therefore, these non-gene areas were selected for genealogy testing, in part so that individuals could compare their results with others without fear of inadvertently disclosing information that some might want to keep private. In addition, these non-gene areas of DNA were often selected for such testing because it was thought that they had no medical relevance, so the testing would avoid the risk of running afoul of a wide variety of ethical and regulatory constraints.

This junk status changed abruptly in September 2012 with the coordinated release of more than two dozen articles in the scientific literature that described for the first time the functions of much of this former "junk DNA."[21]

> A staggering batch of over 30 papers published in *Nature*, *Science*, and other journals this month, firmly rejects the idea that, apart from the 1% of the human genome that codes for proteins, most of our DNA is "junk" that has accumulated over time like some evolutionary flotsam and jetsam.[22]

While this finding has great medical implications, it may not have much impact on genealogical research. These areas formerly considered to be junk apparently do not directly produce our physical characteristics or diseases. However, many of them seem to regulate the genes that do. When properly chosen, they still work well for our family history requirements. They are "junk" no more after having been rehabilitated and upgraded to "noncoding" DNA.

WHAT ARE SINGLE NUCLEOTIDE POLYMORPHISMS?

> Single nucleotide polymorphisms, frequently called SNPs (pronounced "snips"), are the most common type of genetic variation among people. Each SNP represents a difference in a single DNA building block, called a nucleotide.
>
> SNPs occur normally throughout a person's DNA. They occur once in every 300 nucleotides on average, which means there are roughly 10 million SNPs in the human genome. Most commonly, these variations are found in the DNA between genes.
>
> Most SNPs have no effect on health or development. Researchers have found SNPs that may help predict an individual's response to certain drugs, susceptibility to environmental factors such as toxins, and risk of developing particular diseases.[23]

In genetic genealogy research, SNPs are changes in your DNA sequence at specific locations. They can occur anywhere in the DNA. They document junctions in the branching of the flow of human genes from earliest times down to the present. Without them, the DNA of all peoplewould be identical and, therefore, not very informative in helping identify and understand individual differences. Some of the DNA tests of interest to genealogists are based on identifying the most recent SNP. SNP tests are becoming increasingly important as genetic genealogy tests are growing more comprehensive.

WHAT ARE SHORT TANDEM REPEATS?

"DNA is chock full of nucleotide patterns that repeat thousands of time in a single genome. (The sequence GATTACA, which supplied the title of a 1997 science fiction movie, recurs roughly 697,000 times throughout the human genome)," according to science writer Dale Keiger.[24] Note that all the letters are either "A," "C," "G," or."T." While short tandem repeats (STRs) can occur anywhere in our nuclear DNA, so far genetic genealogists have focused on those on the Y chromosome. The number of times that such STRs occur in succession at certain locations on the Y chromosome will be important in the next chapter.

HOW IS DNA TRANSMITTED?

When babies are conceived, the mitochondrion is normally solely determined by the mother. The mother also contributes the recipe for one of the X chromosomes.[25] The father contributes the specifications for one of his sex chromosomes as well. If he contributes a copy of his X chromosome, the child will be a girl. If he contributes a copy of his Y chromosome, the child will be a boy.

Through her egg, the mother contributes a copy of half of her autosomal DNA (atDNA). Through his sperm, the father contributes a copy of half of his autosomal DNA. The resulting offspring inherits half of its DNA from each parent. This part of human inheritance is generally easy to understand.

Where it starts to get complicated is that both of the parents have inherited half of their genomes from each of their parents—the grandparents of the new infant. When it is time for the parents to pass copies of their genomes on to the next generation, only half of it is selected for this transfer. Through a process called recombination, segments along each portion on the genome are chosen. At our present level of understanding, this appears to be a totally random process. We do not understand why one segment from one grandparent's genome is selected instead of the corresponding segment from the other grandparent. Overall, about half of the contribution of each of the grandparents will be passed on to each parent and only half of that will be transmitted to the sperm or the egg. The net result is that the resulting babies will end up with about one fourth of their DNA from each grandparent. However, the amount received from each grandparent will vary from person to person.

When this recombination is replicated to create another child, the random process of selecting which part of the genome of each of the grandparent's DNA is to be inherited is repeated. The result is full siblings who will share only about half of their DNA with each other. This process accounts for the differences of appearance and other characteristics between siblings. And so it goes, from generation to generation, as we investigate our family trees.

FOUR DIFFERENT KINDS OF DNA: A REVIEW

Figure 1.7 provides a concise summary of the different characteristics of the four types of DNA that each of our families have passed down through the generations. At the risk of redundancy, I will repeat key points from this summary throughout the book.

An important caveat to remember is that the results of one kind of DNA test cannot be compared with another. For example, a Y-chromosome test of one person cannot be compared with an autosomal test of another person. This would be somewhat akin to trying find out the distance between Chicago and New York by using a map of Europe. The tests examine different parts of one's DNA and they measure different things.

GETTING STARTED

The first step in DNA testing is to determine your testing goals. There are at least four legitimate motivations for conducting a test of an individual's DNA. The first is

Figure 1.7
Summary of the characteristics of the four types of DNA

DNA	Autosomal	Mitochondrial	X-Chromosome	Y-Chromosome
Who has it?	Everyone	Everyone	Men have 1 chromosome Women have 2 chromosomes	Only men
Who passes it on?	Each parent	Only women to children of either gender	Women to all children; men only to daughters	Only men
Where is it located?	In the nucleus of cells	Inside cells but outside their nucleus	In the nucleus of cells	In the nucleus of cells
How many are passed down?	One copy of each of the 22 pairs of chromosomes in each cell	Multiple copies in each cell	1 chromosome in men and 2 chromosomes in women	1 chromosome to men only
From whom is it inherited?	Half from mother and half from father	All from mother to all children	Men get 1 chromosome from their mother only Women get 1 chromosome from each parent	From father to sons only
Who can be tested for it?	Everyone	Everyone	Everyone (but men have only a maternal contribution)	Men only
What do the most common consumer test results report?	Length of matching segments shared between people	Actual chemical bases; there are only 16,569 locations, so each is important.	Length of matching segments shared between people	Number of short tandem repeats (STRs) at given locations have been the most prominent but SNPs are coming to the forefront in more advanced tests

parallel to our need to preserve the oral stories that our elders heard from their parents and grandparents. We will remember some of them but not with the same precision of detail that they did. Therefore, we need to record them for future generations in some medium more durable than human memory. In like manner, it is important not only for our current research, but also for future researchers of our families, that we preserve the DNA information carried by the oldest generation of our family still living. Although much of that data is inherited by descendants, some of it will be difficult, if not impossible, to reconstruct at a later date. Therefore it is important to preserve that information while it is just a cheek swab or saliva spit away.

Second, many people want to test their DNA to establish a family health history or to take advantage of personalized genomic medicine.

Third, it is also acceptable to test your DNA out of curiosity just to experience the process and learn about it. If this is your goal, anything you learn about your family relationships can be considered a serendipitous bonus.

However, most of us order DNA tests for the fourth reason: because we want to advance our current family history research. You will maximize your chances of success if you focus on a specific question you would like to answer. This will help you determine which DNA test to order and who in your family is most likely to have the answer within the information contained in his or her DNA. You must set your testing priorities based on your answers to the following questions posed by professional genetic genealogist CeCe Moore:

- Are you primarily interested in researching your surname?
- Are there specific brick walls that you wish to target with the use of DNA testing?
- How far back in your family tree are these brick walls?
- What is the ancestral pattern back to these brick walls—for example, mother's mother's mother or father's mother's mother's father?
- Are you ready for a long-term project or do you desire quick answers?
- Are there adoptions in your family tree that you would like to explore?
- Is your primary interest receiving a percentage breakdown of your overall ancestral origins or "ethnicity"?[26]

Your answers to these questions will guide your selection of the most appropriate test and the members of your family who should be tested. Discussions of these testing options and case studies of actual test results will make up most of the balance of this book.

SUMMARY

As you proceed through this book and along your path as a genetic genealogist, remember the differences between the four types of DNA and how each is inherited from parents to children. Here is a quick review:

1. "guy DNA" — a "celibate" or unremixed DNA:
   a. Boys inherit it from their fathers.
   b. Boys do *not* inherit it from their mothers.
   c. Girls do *not* inherit it from either parent.
2. mtDNA — a "celibate" or unremixed DNA:
   a. Boys inherit it from their mothers.
   b. Girls inherit it from their mothers.
   c. Boys do *not* pass it down to their children.
3. atDNA — a "promiscuous" or remixed DNA:
   a. Boys inherit half from their mothers.
   b. Boys inherit half from their fathers.
   c. Girls inherit half from their mothers.

    d. Girls inherit half from their fathers.

    e. Because of recombination:

        i. Siblings do not get the same mixture and on average share only about 50 percent with each other.

        ii. Although grandchildren receive an average of 25 percent from each grandparent, the random nature of inheritance of atDNA makes it impossible to predict in advance what portion will be passed down.

4. xDNA:

    a. Boys inherit it from their mothers.

    b. Girls inherit it from their mothers.

    c. Girls inherit it from their fathers.

    d. Boys do *not* inherit it from their father.

The balance of this book will give examples of how matches between individuals on each of these four types of DNA can be evaluated to extract information relevant to family history researchers.

NOTES

1. Jan Kelly, "DNA Proved My Genealogy to Be Correct!" http://famnet.net.nz/newsletters/famnet/January_2013/DNA%20Proved%20my%20Genealogy%20to%20be%20Correct.pdf, accessed January 31, 2013.

2. Carolyn Abraham, *The Juggler's Children: A Journey into Family, Legend and Genes That Bind Us* (Toronto, ON: Random House Canada, 2013), 5.

3. Angie Bush, communication to the author, February 26, 2014.

4. John Godfrey Saxe, "The Blind Men and the Elephant," http://www.constitution.org/col/blind_men.htm, accessed January 14, 2013.

5. U.S. National Library of Medicine (NLM), National Institutes of Health (NIH), *Genetics Home Reference: Your Guide to Understanding Conditions,* http://ghr.nlm.nih.gov.

6. NIH, "What Is a Genome?" *Genetics Home Reference: Your Guide to Understanding Conditions,* http://ghr.nlm.nih.gov/handbook/hgp/genome, accessed January 16, 2013.

7. Centre for Genetics Education, "Genes and Chromosomes: The Genome, Fact Sheet 1," http://www.genetics.edu.au/Information/Genetics-Fact-Sheets/Genes-and-Chromosomes-FS1/view, accessed January 10, 2013.

8. National Institutes of Health, "What Is DNA?", http://ghr.nlm.nih.gov/handbook/basics/dna, accessed January 10, 2013.

9. National Institutes of Health, "What Is Mitochondrial DNA?", http://ghr.nlm.nih.gov/handbook/basics/mtdna, accessed April 7, 2014.

10. National Institutes of Health, http://ghr.nlm.nih.gov/handbook/illustrations/cellparts?show=cellmitochondria, accessed January 10, 2013.

11. National Institutes of Health, "What Is a Chromosome?", http://ghr.nlm.nih.gov/handbook/basics/chromosome, accessed January 11, 2013.

12. National Institutes of Health, "Chromosomes 1-22, X, and Y", http://ghr.nlm.nih.gov/handbook/illustrations/chromosomes, viewed 3/14/14.

13. National Institutes of Health, "How Many Chromosomes Do People Have?", http://ghr.nlm.nih.gov/handbook/basics/howmanychromosomes, accessed January 11, 2013.

14. National Institutes of Health, "X Chromosome," http://ghr.nlm.nih.gov/chromosome/X, accessed January 11, 2013.

15. National Institutes of Health, "What Is the Y Chromosome?", http://ghr.nlm.nih.gov/chromosome/Y, accessed January 15, 2013.

16. National Institutes of Health, "What Is a Gene?", http://ghr.nlm.nih.gov/handbook/basics/gene, accessed January 11, 2013.

17. National Institutes of Health, "What Is a Gene?"

18. Bryan Sykes, *DNA USA: A Genetic Portrait of America* (New York, NY: Liveright, 2012), 18.

19. Dale Keiger, "The DNA Data Flood," *Johns Hopkins Magazine* 65 no. 3 (Fall 2013): 21.

20. Michael C. Schatz and Ben Langmead, "The DNA Data Deluge: Fast, Efficient Genome Sequencing Machines Are Spewing out More Data Than Geneticists Can Analyze," *IEEE Spectrum* (June 27, 2013), http://spectrum.ieee.org/biomedical/devices/the-dna-data-deluge, accessed October 30, 2013.

21. Kent Sepkowitz, "The Inner Life of Cells: DNA's Middle Managers Could Be the Key to Future Cures," *Newsweek.* (September 17, 2012): 8.

22. "Junk DNA Not Junk after All," *Medical News Today* (September 9, 2012), http://www.medicalnewstoday.com/articles/250006.php, accessed October 6, 2012.

23. Centre for Genetics Education, http://ghr.nlm.nih.gov/handbook/genomicresearch/snp, accessed January 16, 2013.

24. Keiger, "The DNA Data Flood," 22.

25. The inheritance pattern of X chromosomes is somewhat complex and will be discussed in more detail later in this book.

26. CeCe Moore, "DNA Testing for Genealogy: Getting Started, Part One," *Geni* (July 18, 2012), http://www.geni.com/blog/dna-testing-for-genealogy-getting-started-part-one-375984.html, accessed December 13, 2013.

2

Who Is the Father? "guY" DNA

DNA testing of potential interest to genealogists exists in several separate and distinct forms. Some of them, with which you already may be familiar, are as follows:

1. Medical testing for specific genes which have an impact on various health risks and our bodies' reactions to various potential pharmacologic remedies. This is particularly useful in documenting family health histories when certain health risks are said to "run in the family."
2. Parental testing to establish the father and/or mother of a specific individual can be a potent tool for adoptees.
3. Haplogroup testing for deep ancestry which have anthropological value in identifying the migration pattern of one's ancient clan or tribe back long before humans had written records.
4. Genealogical testing which establish or eliminate possible family relationships or discover previously unsuspected connections.

The latter two types of testing are discussed in detail in this book.

Using DNA to trace family history is a concept that was introduced to the general public on the eve of the 21st century.

> On 5 November 1998 the journal *Nature* placed an inaccurate and misleading headline based on this study which read, "Jefferson Fathered Slave's Last Child." Most of the mass media and many others assumed the headline to be correct. At the time Daniel P. Jordan, Ph.D. and President of the Thomas Jefferson Memorial Foundation (TJMF), stated that "Dr. Foster's DNA evidence indicates a sexual relationship between Thomas Jefferson and Sally Hemings." Subsequently Mr. Jordan admitted that "after the initial rush to conclusions came another round of articles explaining that the study's results were less conclusive than had earlier been reported."[1]

At no time was there any indication that the science behind this pioneering study was wrong. However, at times the interpretation was extended a little further than the evidence would support. Even though the headline as written was probably true, we need to be careful not to claim conclusions more encompassing than our evidence will support.

Geneticist and genealogist Angie Bush helps us understand the place that DNA results have in our overall research process:

> In this particular case (as with most cases) it was the combination of the DNA results with historical documentation that allowed researchers to determine that Thomas Jefferson fathered Sally Hemings' child. In most cases, DNA is not some type of "trump card," nor does it provide a magical answer. It must be used as another "record type" and as part of the genealogical proof standard. Often neither documentary nor DNA evidence alone allows conclusions regarding kinship and identity to be reached—all pieces of evidence must be correlated together to reach satisfactory conclusions.[2]

WHO'S THE FATHER?

The question asked in the Jefferson-Hemings project is one of the first questions asked by anyone doing family history research: "Who's the father?" This question is repeated over and over again as new family units are added to the research project. The answers to this question become the information entered into the top boxes of the emerging pedigree chart for each succeeding generation.

Most of us carry an important clue to this question's answer. At least for the first generation of our research, it is our surname—or at least it is the surname assigned to us when we were born. Western European tradition for some of us has prescribed this custom for the last several centuries. Scandinavian cultures have followed this tradition for more than a century. African Americans have had surnames since the Civil War, and those who earned their freedom earlier have carried them even longer. Unless we were adopted or were assigned our mother's surname at birth, we start our research with this important clue about the identity of our biological father.

Surname studies have long offered a greater fascination to Europeans than they have to those of us in North America. Many European surnames were derived from place names. In addition, surnames came into existence on that continent during much less mobile times. As a consequence, they flourished in specific localities on the eastern side of the Atlantic.[3] By the time of the European settlement of North America, the vast majority of the colonists already had surnames. In addition, the persistent westward migration of European settlers further weakened the association of surnames with specific locations. As a result, surnames have not been attached to other regional locations to the same extent as they have been in Europe.

In addition to our surname, most of us have family oral traditions and official and other documents to help us answer the father question. These traditional sources for genealogists remain very important today.

All of us carry important clues to our paternity in our cells. Each of us received half of our autosomal DNA from our biological fathers. Autosomal DNA will be discussed in detail in subsequent chapters. Males also carry a more specific clue—the DNA in their Y chromosomes. That chromosome is inherited by all males directly from their father and it is identical or nearly identical with that of the father.

Y-chromosome tests explore the paternal ancestral line shown in the dark boxes of Figure 2.1. They probe deep ancestry along that line far beyond the generations shown in this simplified illustration.

Figure 2.1
Y-chromosome tests explore the paternal ancestral line

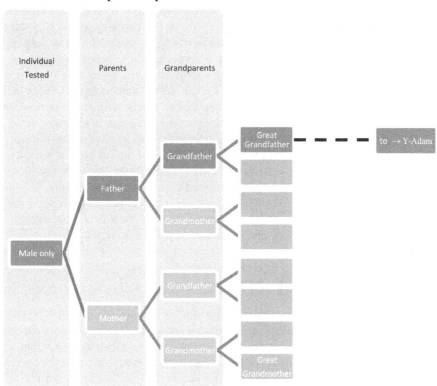

EARLY "GENETIC GENEALOGY" TESTING

Direct-to-consumer (DTC) Y-chromosome testing began to be used by genealogists in 2000. Some very small projects, like the Jefferson-Hemings study mentioned earlier, existed before then. However, the laboratory analyses for these small private studies were carried out in academic labs whose scientists are guided by the "publish or perish" paradigm to focus their efforts on projects that will lead to new knowledge. Generally they do not want to spend time and energy on commercialization of their discoveries and inventions. This factor explains why we see the divisions of labor such as those between Mike Hammer and Bennett Greenspan. Hammer, at the University of Arizona, concentrated on further discoveries, while Greenspan made these finds and others available to consumers.

In the late 1990s the first draft of the human genome was far from complete, but in isolated labs DNA was being used in ways that would soon be useful to genealogists. Other studies contemporaneous with the Jefferson-Hemings project resulted in scientific papers, a few of which caught the attention of the mass media. This attention caused some members of the general public to start approaching geneticists at Oxford University and at the University of Arizona with requests that their family's DNA be tested. In England, geneticists, including Bryan Sykes (who was soon to write *The Seven Daughters of Eve*), set up an independent company to separate the testing process for genealogists from the ongoing academic genetic research. Thus was born the consumer-oriented DNA testing operation Oxford Ancestors.

On the other side of the Atlantic, a testing company called Family Tree DNA (FTDNA) was created in Houston by entrepreneur Bennett Greenspan. Bennett has been a lifelong genealogist. When he read in the *Wall Street Journal* about genetic research being conducted by Dr. Michael Hammer of the University of Arizona, he immediately thought of a genealogical question related to his mother's family that DNA might answer. In his attempts to get samples tested that might answer his family history question, he was rebuffed by Hammer. In their discussions, Hammer mentioned that he had been approached by others seeking such testing, but he did not want to get into the commercial side of consumer genetic testing. Subsequently, in 2000 Greenspan set up a consumer-oriented company to provide DNA testing services for genealogists who had family history questions similar to his own. Hammer became a scientific consultant but was not directly involved with the operation of the company.

Thus, on both sides of the Atlantic, the dawn of the new century offered a new tool to genealogists. Now we as genealogists had much to learn about how properly to apply the data that genetics could provide. Y-chromosome testing was going to take the lead for much of the next decade.

SORENSON MOLECULAR GENEALOGY FOUNDATION

As early as 1999, billionaire philanthropist James Sorenson dreamed of "creating a genetic map of the peoples of the world that shows relationships shared by the entire human family."[4] This would be accomplished by creating DNA databases that were connected to pedigree charts. After several smaller projects were pursued, the Sorenson Molecular Genealogy Foundation (SMGF) was formed in 2002 as a not-for-profit organization. By the time of Sorenson's death in 2008, SMGF had collected "more than 100,000 DNA samples, together with four-generation pedigree charts, from individuals in more than 100 countries around the world."[5] The resulting Y-chromosome DNA and mitochondrial DNA databases are still searchable. A projected autosomal database was never launched.

After Sorenson's death, SMGF lost both its funding and its forward momentum. This led to the association of the foundation with the genetic genealogy company GeneTree. The subsequent acquisition of both of these entities by Ancestry.com has been chronicled by genetic genealogist extraordinaire Tim Janzen, MD.[6] Ancestry's plans for the future of this resource remain unclear.

FINDING Y-CHROMOSOME INFORMATION ABOUT
YOUR PATERNAL LINE *BEFORE* YOU TEST

You may be able to learn something about the Y-chromosome DNA of your paternal line even before you get test results back from the lab on yourself or on a close male relative. At least two avenues for exploration are available before you (or a close male relative) submit a sample of saliva to the lab.

The first avenue of exploration is to review the results of the DNA surname project (if there is one) for your particular surname of interest. Most surname projects are hosted by FTDNA and/or run by volunteer administrators through FTDNA. Some surname administrators have set up separate websites that contain results from several

Figure 2.2

Sample report from a surname project results page

Kit Number	Name	Paternal Ancestor Name	Haplogroup	DYS393	DYS390	DYS19	DYS391	DYS385	DYS426	DYS388	DYS439	DYS389i	DYS392	DYS389ii	DYS458	DYS459	DYS455	DYS454	DYS447	DYS437	DYS448
A - Virginia Group 1																					
MIN				13	25	14	11	11-13	12	12	12	13	14	29	16	9-10	11	11	24	14	18
MAX				13	25	14	11	11-13	12	12	12	13	14	29	16	9-10	11	11	25	15	18
MODE				13	25	14	11	11-13	12	12	12	13	14	29	16	9-10	11	11	25	15	18
97346	Dowell	Issac Dowell b. 2 Nov 1826, Ohio, USA	R-M269	13	25	14	11	11-13	12	12	12	13	14	29							
110434	Bartley	Vernon Creasy Bartley	R-M269	13	25	14	11	11-13	12	12	12	13	14	29	16	9-10	11	11	24	15	18
48335	Bartley	John Dowell b. ca. 1700 VA, USA	R-M269	13	25	14	11	11-13	12	12	12	13	14	29	16	9-10	11	11	25	14	18
18823	Bartley	John Dowell b. ca. 1700 VA, USA	R-M269	13	25	14	11	11-13	12	12	12	13	14	29	16	9-10	11	11	25	14	18
120732	Dixon	Samuel Roosevelt Dixon b.1904 NC d.1964 In.	R-M222	13	25	14	11	11-13	12	12	12	13	14	29	16	9-10	11	11	25	15	18
62586	Dowell	Reuben Dowell, ca. 1805 Tennessee, USA	R-M269	13	25	14	11	11-13	12	12	12	13	14	29	16	9-10	11	11	25	15	18
53095	Dowell	Nehemiah Dowell, 1733, Prince William County, VA	R-M222	13	25	14	11	11-13	12	12	12	13	14	29	16	9-10	11	11	25	15	18
95431	Dowell	Reuben Dowell, born ca. 1805 Tennessee, USA	R-M269	13	25	14	11	11-13	12	12	12	13	14	29	16	9-10	11	11	25	15	18
117449	Dowell	John Dowell, 1730, Prince William County, VA	R-DF23	13	25	14	11	11-13	12	12	12	13	14	29	16	9-10	11	11	25	15	18
108864	Dowell		R-M222	13	25	14	11	11-13	12	12	12	13	14	29	16	9-10	11	11	25	15	18
6007	Bartley	John Dowell b. ca. 1700 VA, USA	R-M222	13	25	14	11	11-13	12	12	12	13	14	29	16	9-10	11	11	25	15	18
63842	Dowell	Nehemiah Dowell, 1733, Prince William County, VA	R-M222	13	25	14	11	11-13	12	12	12	13	14	29	16	9-10	11	11	25	15	18
118721	Bartley	Vernon Creasy Bartley	R-M269	13	25	14	11	11-13	12	12	12	13	14	29	16	9-10	11	11	25	15	18
160095	Dowell	Nehemiah Dowell, 1733, Prince William County, VA	R-P311	13	25	14	11	11-13	12	12	12	13	14	29	16	9-10	11	11	26	15	18
79748	Dowell	Thomas Dowell ca. 1705 Prince William Co., VA	R-M222	13	25	14	11	11-13	12	12	12	13	14	29	16	9-10	11	11	26	15	18
24985	Dowell	John Dowell b. ca. 1700 VA, USA	R-M222	13	25	14	11	11-13	12	12	12	13	14	29	16	9-10	11	11	26	15	18

different companies and display their information in a slightly different manner than the FTDNA-hosted projects. You can "Google" your particular surname project and/or get to it directly through the Family Tree DNA website (www.ftdna.com). This is the lab with the largest database of tested males. When you arrive at the FTDNA home page, follow these steps:

1. Enter the surname of interest in the "Search Your Surname" box on the right as you scroll down the page.
2. Click in the most appropriate link(s) under "Projects" on the left side of the page.
3. Click on the "Website" in the upper-left corner of the page.
4. Click on the results link. This will probably be part of a banner across the top of the page, but it may be in the left column or elsewhere depending on the "creativity" of the designer of the website. Ideally, the link will lead to an Excel-like table that includes a column with a heading like "Most Distant Ancestor" or "Paternal Ancestor Name" as in the truncated table shown in Figure 2.2.
5. Also look on the first project page to see if pedigree lines of participants have been posted or linked.

The chart in Figure 2.2 is simplified and abbreviated for purposes of illustration. A "live" surname page probably would show "**DNA Y**-Chromosome **S**egment Number" (DYS) columns stretching out to the right to display up to 111 markers for those members who have been tested to the current maximum. The number of participants may be several hundred in the case of common surnames. In that case the number of rows displayed would extend far down the page and may be displayed on multiple pages. You may benefit from using your browser's "Find" command to help you zero in on a particular name or location. To the extent possible, the volunteer administrator(s) for that project will have grouped the members with other members who seem to be closely related on the basis of common ancestors, common locations, and/or similar test results. Figure 2.2 displays a subgrouping of a much larger project. At the beginning of each subgroup, the chart may display the minimum DYS value, the maximum DYS value, and the most commonly appearing (mode) value for that particular subgrouping.

Also displayed will be a probable haplogroup for each member. More will be explained about haplogroups later in this book. For now, it is sufficient to know that men closely related within recent centuries will be members of the same haplogroup. Men who are not members of the same haplogroup are not closely related. In Figure 2.2, one member has a much longer haplogroup designation than the others. If you are looking at a live webpage displaying such data, his haplogroup (R1b1a2a1a1b4b, in this case) will generally be displayed in green letters. This designation indicates that his haplogroup has actually been tested and confirmed in the lab. The others in this particular group are shown to belong to a stub or truncated version R1b1a2. They have not been specifically tested to determine their haplogroup membership. Instead, they have been estimated to belong to this "backbone" of the haplogroup based on the similarity of their results to the results of others who have been tested. In the case shown in Figure 2.2, it is highly likely that all—or certainly most—of these men would be found to share the much more precise designation if they were tested specifically for haplogroup membership. However, the 12-marker results shown would not be sufficient to safely make this assumption. Tests of 37 or more markers and other genealogical information have allowed the project administrator to make this grouping. Although two different surnames are included, genealogical information allowed the project administrator to create the subgroup with confidence.

If information about the earliest ancestor is not included, look for an email address for the project administrator or a "Contact Us" link. Ask the administrator if any of the participants are descended from an ancestor who originates from the ancestral home of your line. (Example: "Do any of the Fishers in your project originate from Syleham, Suffolk, England?")

If you are fortunate enough to find an obvious match to your documented ancestral line, at a minimum you should be able to find a general haplogroup for your line. You may also discover some typical marker values for your ancestral line. Interpret this information with care: some of the actual marker values of men who descend from the same common ancestor will vary somewhat over the generations. Likewise, you may find some genealogical information as a bonus. You may also find a knowledgeable project coordinator who can be helpful to you. However, do not become a pest. These accommodating project administrators are volunteers who often have day jobs and lives outside the fascinating world of genetic genealogy.

The second avenue of exploration you can pursue prior to receiving your test results from the lab is to search by surname at two other databases: SMGF (www.smgf.org) or Ysearch (www.ysearch.org). The latter is a site on a server operated by FTDNA. It is up to the initiative of an individual to enter his results here. Although it is easy to transfer results from FTDNA to Ysearch automatically, do not assume that most of those persons tested at FTDNA have done so. Those tested at other labs must enter results manually.

To find if anyone of your surname and location has results at Ysearch, enter the website and follow these steps:

1. Select the "Search by Last Name" tab.
2. Enter the surname of interest in the top search box. Note that you need to enter the words you see at the bottom of the page in the appropriate box to prevent data mining by automated programs. Some of the words given are seriously distorted and difficult to make out. You may have to click on the reload button to the right of the box to request another set of words if you have difficulty reading those initially presented.
3. Click on the "Search" button and wait patiently. If the "Search" button grays out, the program is searching. Although searching large databases is getting faster all the time, searching DNA

Figure 2.3
Sample results from an Ysearch query

Last Names Matching "Dowell"

Search by Last Name > Search by Last Name Results > Last Names Matching "Dowell"

33 user(s) found.

Check the boxes of the individuals you want to compare and then click the underlined word "COMPARE" at the top of the column

Check All - Clear All

Compare	User ID	Pedigree	Last Name	Ancestor	Origin	Haplogroup	Tested With
☐	JDUMZ		Dewell	Dewell	Unknown	R1b1a2 (tested)	Family Tree DNA
☐	3QHGD		Dowell	Dowell	Maryland, USA	Unknown	Family Tree DNA
☐	42VVY		Dowell	Dowell	Maryland, USA	Unknown	Family Tree DNA
☐	4ESJN		Dowell	Dowell	Unknown	Unknown	Family Tree DNA
☐	4R4UW		Dowell	Dowell	Calvert County, Maryland, USA	Unknown	Family Tree DNA
☐	5ZXF7		Dowell	Dowell	Unknown	R1b1a2 (tested)	Family Tree DNA
☐	6DW5T		Dowell	Dowell	England	Unknown	Family Tree DNA
☐	7X2BD		Dowell	Dowell	Unknown	R1b1a2a1a1b4b (tested)	Family Tree DNA
☐	8AUB9		Dowell	Dowell	Fauquier County, Virginia, USA	Unknown	Family Tree DNA
☐	AWX7M		Dowell	Dowell	Calvert and Anne Arundel counties, Maryland, USA	Unknown	Family Tree DNA
					Greenfield, Orange Co., Virginia		

results for thousands of individuals can take several seconds or more, particularly if many searches are being conducted simultaneously.

4. Carefully examine the spelling of surnames and origins shown in your results page similar to the example given in Figure 2.3.

5. Click on the "User ID" for any entry that looks promising. Be sure to scroll to the bottom of the page. Often this will lead to notes that have been entered by the person who posted that entry.

6. If the entry still looks promising, click on "Contact this user." This will allow you to send a blind email to the listing individual. Be sure to enter the code given in the bottom box below your message.

7. Return to the results page you got after Step 3. See if any entry of interest has "Show" in the Pedigree column. This indicates that a GEDCOM file pedigree has been uploaded. If you do not see the desired surname, look for a red arrow on the right margin that is pointing to a continuation of the chart on another page.

8. If you are interested in looking for surnames that are not on your direct paternal line, you can return to the results page after Step 2. Click on the number in the box under "Pedigree." This will display all instances of your surname of interest in uploaded GEDCOMs, no matter whether they are on the paternal line or not. Follow the procedure for either Step 5 or 7 if appropriate.

The following steps will allow you to search the SMGF database by surname:

1. In your browser, go to www.smgf.org.
2. At the bottom of the left column, select "Y-Database."
3. From the left column, select "Search the Y-Database."
4. Enter your login information or click on "Sign up now" for a free account.
5. Select "Perform a new search" at the bottom of the Options list.
6. Select "Search by Surname" in the Parameters box.
7. Click on "Search" toward the bottom of the box.
8. Enter the randomly generated number in the box below it and click "Submit."

These preliminary explorations will not take the place of actual testing. You will never be sure your paternal line is a direct connection unless you have test results from an individual to whom you know you are closely related. However, these steps will give you something to do during the weeks you are impatiently waiting for results to come back from the lab.

WHAT CAN YOU EXPECT TO LEARN FROM A yDNA TEST?

Just as you don't need to be an aeronautical engineer to be able to fly in an airplane, you don't need to have a degree in genetics to be able to interpret the results of a DNA test for genealogical purposes. It's not the data points themselves that are important. Rather, the meaning of the test results lies in how your results compare with those of others. On genetic tests for genealogical purposes, there are no "right" or "wrong" answers. Everything is about with whom you cluster and with whom you do not.

Lab results for an entry level Y-chromosome test might look like the sample shown in Figure 2.4. The row for "Locus" in these results is the numerical sequence of the markers being examined. For example, Locus 1, or the first result reported, is for *DNA Y-chromosome segment number (DYS#)* 393—a specific short tandem repeat (STR) on the Y chromosome.[7] While DYS# 393 may be very exciting to a geneticist, the only significance it has for you and me is that it has a value of 13, which can be compared with the value other men have for that same DYS#. The value of 13 has no known significance of its own. It is not better or worse than a value of 12 or 14, for example; it is just the first result for the tested person. The lab doing the analysis in this instance was FTDNA. There is increasing standardization among the reporting schemes from the various labs that analyze DNA.

DNA test results for an individual are meaningless in isolation. They take on meaning only when they are compared with databases containing the results of many people. In this case, big is definitely better. The larger the database against which you can compare your results, the more likely you will be to learn something useful about your family history. You also want to fish in as many ponds as possible. Remember that anyone who has been tested at any lab can enter Y-chromosome results here at no charge to find matches between individuals who have tested at different labs.

Figure 2.4

Sample lab report of the results of a 12-marker yDNA test

PANEL 1 (1-12)											
Marker	DYS393	DYS390	DYS19**	DYS391	DYS385	DYS426	DYS388	DYS439	DYS389I	DYS392	DYS389II*
Value	13	25	14	11	11-14	12	12	11	13	13	29

The actual areas of DNA used for Y-chromosome analysis are from parts of human DNA that have long been called "junk DNA." Until very recently, these particular parts of our DNA were thought to have no purpose in the passing of characteristics from one generation to the next, but new research is showing that they may have some function in turning genes "on" or "off," as well as possibly other functions. This topic will not be discussed here in depth; just be aware that "junk DNA" is not truly "junk." These DYS# numbers are not for locations along our genomes that determine whether a person will have blue eyes or brown eyes, be tall or short, or be likely to get breast cancer. Therefore, they were selected for analysis in part so that individuals could compare their results with others without fear of inadvertently disclosing information that some might want to keep private.

Labs testing of human Y chromosomes for genealogical purposes has sought to strike a balance between "fast-mutating markers" and "slow-mutating markers." This balance is important because we need some indicators that have changed in "genealogical time." "Genealogical time" refers to the last 400 years or so during which most people have had surnames and some form of written documentation for some events of our lives. These mutations are good for family historians who use Y-chromosome tests because they allow us to differentiate between lines of ascent backward in time. At the same time, there must be some markers that generally stay the same from one generation to the next so that we have a trail to follow.

Some genetic genealogists want to analyze every marker on which different values are reported for supposedly related individuals to determine whether these differences occurred on "slow"- or "fast"-changing markers. This point may be important when studying large populations. It can lead to learning about the causes and possible effects of such mutation patterns. However, for most family history purposes today, "a mutation is a mutation." Once it occurs, we expect it to be passed on to male offspring.

Taken in combination, the Y-chromosome values can indicate whether two individuals are likely to have a common male ancestor within the genealogical era, the period in which most families have had surnames. Therefore, it is by comparing these results with others that they take on value. Such a comparison can help two individuals, or a group of individuals, determine whether they are closely related. If two individuals have the same exact 12-marker values *and* the same (or a close derivative) surname, this probably would indicate that they had shared a common male ancestor sometime within the last few hundred years. If there is a single mismatch and a similar surname, a close relationship is also possible.

However, when we compare Y-chromosome DNA, we are dealing with probabilities and not absolute certainties. In this instance, the *CSI*-type television shows have done us a disservice by leading us to believe that all DNA matches are absolute. Y-chromosome tests are not this precise.

A 12-marker test is very minimal. Several mismatches would rule out a close familial match. Note in Figure 2.5 that the two males of the same surname have mismatches at locations 6, 9, and 11. This is called a 9/12 match (i.e., 9 matching markers out of 12 possible matches). It might also be said that the two men have a genetic distance (gd) of 3 because each mismatch is different by one unit.

Although these two men had done extensive research on their surname and had long assumed that they had a common male ancestor, perhaps in Colonial times, this result suggests this was not correct. Subsequent extensions of these tests confirmed that they were not closely related. In fact, their closest common paternal ancestor probably lived between 3,000 and 3,500 years ago—long before surnames were in use. In most cases,

Figure 2.5

The 12-marker results from two men with the same surname

Locus	1	2	3	4	5	6	7	8	9	10	11	12
DYS #	393	390	19	391	385a	385b	426	388	429	389-1	392	389-2
Man 1	13	25	14	11	11	**14**	12	12	**11**	13	**13**	29
Man 2	13	25	14	11	11	**13**	12	12	**12**	13	**14**	29

at least 37 markers would be needed to have any confidence in confirming a match. At that level the two men in this example had a genetic distance of 18. Clearly, they are not closely related on their surname line.

We have no magic way to determine in advance how many Y-STR markers a man should test. FTDNA, the testing leader, has traditionally offered 12-, 25-, 37-, 67-, or 111-marker tests to new customers. At the moment, the 25-marker tests for new customers are available only through a surname project. Generally, 37 markers is a good starting level. However, a few men will have no matches or only a couple of matches at 12 markers and none at a higher level of resolution. It all depends on how well the region of origin of one's ancestors has been sampled and whether that person's paternal line has been decimated by factors like war, disease, or genetic disorders. By comparison, for men from very prolific paternal stock and heavily sampled regions, there may be 1,000 matches at 12 markers, half of them exact. For them, at least 37 markers and a similar surname are necessary to have any confidence regarding a relative who shares a common male ancestor within the last several generations.

Such DNA information, taken in isolation, is not nearly as useful as it can become when it is combined with the results of traditional genealogical research. In this regard, parallel reconstructions can sometimes be useful. Back in 2005, it became apparent that several Dowells have very close DNA matches with several men whose surname is McDaniel. Based on the early Y-chromosome DNA evidence that was available at that time, there was a statistical probability of almost two chances in three that these Dowells and McDaniels all shared a common male ancestor within the last eight generations. This is approximately the length of time each group has established that their ancestors have been on the western side of the Atlantic.

When these results were received from the lab, the two groups began to exchange pedigree charts and other genealogical information to explore where their respective lines might intersect. Could there have been an adoption, a name change, an out-of-wedlock child, or some other form of "non-paternity event" (NPE)? No obvious explanations emerged.

As mentioned above two males who share the same surname, but different haplogroups, of necessity have acquired that surname in different manners, locations, legal means, and so on. In other words, the name grew up separately in two or more locations. For example, the name Smith grew up in many locations because most villages had a blacksmith at the time when surnames were being adopted. His family often became known as the Smiths. The acronym NPE is used in both genetic genealogy and clinical genetics as a short form for a non-paternity event. Genetic genealogist Georgia Boop describes its use:

> In genetic genealogy the term NPE is often used in a wider context to indicate a break
> in the link between the Y-chromosome and the surname. Such a breakage may occur

because of formal or informal adoption, illegitimacy inside ("extramarital event"/infi-
delity or rape) or outside of marriage, child known by other surname (mother's maiden
name, stepfather's name), the use of an alias or a deliberate change of surname.[8]

In the case of the McDaniel and Dowell families, no evidence of an NPE has been
discovered. As more men of each surname were tested and their results extended to
67 markers, it became possible to establish that each group had an early ancestor with
exactly the same 67-marker DNA signature as some of their current-day descendants.
In other words, the two North American founding ancestors of these two clusters were
also separated by a genetic distance of 3, just as are several of their current-day descendants. The two lines were equally as far apart genetically in the 18th century as they are
today.

Although there is a very high probability that members of both of these two distinct
groups all share a common male ancestor, that forefather is much further back—almost
certainly not in North America. In addition, they have established, through the triangulation of the 67-marker Y-chromosome DNA results of living Dowell distant cousins,
that some of them have no mutations from the 67 DNA markers of their Dowell
common ancestor who died early in the 18th century. Likewise, for some living
McDaniel distant cousins, their 67-marker Y-chromosome DNA results do not vary
from those of their common McDaniel ancestor who also lived in the 18th century.[9]
That means the early ancestors of both these Dowells and McDaniels, who were in
North America almost 300 years ago, were the same distance apart in their 67 DNA
markers as some of their descendants living today. Therefore, when we take this information into account, we must adjust the probability timeline back to when each group
was likely to have had a common ancestor. When the additional information is included
that there was no common ancestor for either group in at least the last eight generations,
the projected probabilities are adjusted as shown in Figure 2.6.

Probabilities when only DNA results are considered compared to probabilities when
DNA and family history research are combined show a diverging projection in recent
generations but come together further back in time. The probabilities shown in Figure
2.6 are based on 67-marker comparisons and were calculated by TiP.[10] After this

Figure 2.6
More accurate probability models can be constructed when DNA and family history
research are combined

	Likelihood of a Common Male Ancestor	
Generations Back to a Common Male Ancestor	**Based Only on DNA Results**	**Combining DNA with Family History Information**
4	22.22%	NA
8	66.55%	22.42%
12	90.00%	76.81%
16	97.58%	94.38%
20	99.48%	98.81%
24	99.90%	99.77%

adjustment is made, a very good probability (94 percent) emerges that the two groups share a common male ancestor within 16 generations. There is a better than a coin-flip probability (50 percent) that the match occurred within the last 11 or 12 generations. A common male ancestor becomes almost a statistical certainty, well in excess of a 99 percent probability, within 24 generations. However, that far back—600 to 800 years ago—surnames were not in common use, and North America was inhabited only by Native Americans. These two groups do share a common male ancestor, but he lived further back in time than was originally thought when the STR DNA test results were first viewed.

More recently, 111-marker tests have added support to that conclusion. With these results, the unadjusted probabilities of a common ancestor within 8 generations and 12 generations dropped to 26 percent and 66 percent, respectively. When the information was added that there was no common ancestor within eight generations, the probabilities dropped to 11 percent and 59 percent.[11] The common male ancestor does not appear to have lived in North America. As this book was making its way through the editorial process in 2014, BIG Y SNP testing results confirmed that my Dowells and these McDaniels share a common male ancestor but suggest he may have lived several hundred years ago. We hope to refine this estimate as more men are SNP tested. The differences between STR testing and SNP testing of yDNA data will be discussed later in this chapter.

yDNA SURNAME PROJECTS: A CASE STUDY

During the first decade of the 21st century, the most productive application of DNA testing to genealogy was in Y-chromosome surname projects. British surname study expert Chris Pomery described them as follows: "Recently, the development of Y-chromosome DNA tests has created a process so that men, sharing the same surname, can verify whether their previous documentary research has assigned them to the correct tree or direct their future research toward documenting it."[12] This is a process the Dowells initiated in 2004 when the first ones began DNA testing.

The Dowell DNA project started—as any good DNA testing process should—as an effort to test genealogical hypotheses.[13] Initially four independent hypotheses were proposed by project participants to explain how various subgroups of the Dowells were related:

1. The Dowells descended from the McDowells, as was claimed by various oral traditions.
2. A Bartley woman conceived children with a Dowell male whom she never married. She gave the children her maiden name, which her descendants carried to the present day according to their family's oral tradition.
3. The Maryland Dowells and the Virginia Dowells were closely related.
4. The Dowells descended from the Maryland Dowells.

If you think about it, Hypothesis 1—that the Dowells descended from the McDowells—is counterintuitive. "Mc" literally means "son of." Logically the McDowells should have been expected to have descended from the Dowells. However, almost every branch of Dowells has passed down an oral tradition of descent from McDowells. They were said to originally have been called McDowell but at some point the *Mc* was dropped. Therefore the first several Dowells (and Bartleys) to be DNA tested did so within the McDowell surname project. DNA results soon put to a merciful death this mental exercise of which came first, the "Mc" or the non-"Mc": no close matches between any Dowell DNA and any McDowell DNA were forthcoming. That was the first nugget of new learning from DNA testing in this surname project. A separate Dowell project soon evolved.

In contrast, the testing of Hypothesis 2—the Bartleys descending from a male Dowell—was confirmed by testing. The first Dowell and Bartley to be compared matched on 24 of the 25 markers. This gave them a 59 percent probability of having a common male ancestor within 8 generations and a 79 percent probability within 12 generations.[14] As more men were tested, the two who emerged as the best representatives of the Virginia Dowells and Bartleys turned out to be one Dowell and one Bartley who matched exactly on 67 markers. This gave them a 99 percent probability of having a common male ancestor within 8 generations. The results confirmed the oral tradition that a Bartley woman had children with a Dowell man; they had never married, and she gave her children the surname Bartley, which her descendants still carry a couple of hundred years later.

The next finding was jolting to some long-time Dowell surname researchers. Even a year or two after receiving the conclusive DNA results, some otherwise rational and objective genealogists were still in a state of disbelief. It long had been assumed that all Dowells were related in some way. If they just were able to push the paper trail back one more generation to the immigrant, they believed, he would turn out to the "missing link" between his descendants who settled in both Maryland and Virginia. There were some Dowell researchers who could trace their ancestry back to early Colonial Virginia and some who could trace their roots back to early Maryland. No big deal. Back then, people traveled by water when they could. Roads were barely passable at best since President Eisenhower was not born yet and had not yet built the interstate highway system. It appeared obvious that upon passing through Hampton Roads, some Dowells had gone west up the James River in Virginia and others had turned right and gone north up the Chesapeake Bay into Maryland. These were the paths that goods took when they came from England and were followed in reverse when tobacco was exported. It would have been a relatively easy trip from central Virginia to southern Maryland via these waterways. But it did not happen like that. DNA results have established that the Maryland Dowells and the Virginia Dowells have not had a common male ancestor for about three millennia—a period that would extend to a time far earlier than when surnames like Dowell began to be used. The use of this name had emerged independently in at least two separate locations.

Another surprise to experienced researchers was that the Y-chromosome pattern of one small branch of the supposed Virginia Dowells was genetically identical to the DNA pattern of group of Martins rather than to the main group of Virginia Dowells. These Dowells, subsequently christened the Virginia Group 2 Dowells, within the project were called the "Martinized Dowells" by one researcher. The Virginia Group 2 Dowells and their matching Martins can be traced back to ancestors who lived on adjoining properties in mid-18th-century Virginia. Now they are trying to figure out a way to determine whether these folks who have carried the surname Dowell since at least a generation prior to the Revolutionary War could be connected on their maternal side of the family to the main group of Virginia Dowells, who are now known as the Virginia Group 1 Dowells.

And then there is the Dewell line. Printed genealogies for at least the last 80 years have attached the Dewell line of Maryland and Ohio to a branch of the Maryland

Dowells. The Dewells were thought to have changed the first vowel of their names from "o" to "e" somewhere along the line. The National Society of the Daughters of the American Revolution (DAR) has granted membership to some Dewells based on their alleged descent from a Maryland Dowell who had provided service during the Revolutionary War.[15]

DNA tests of Dewell men show them to be related to each other. However, they are much further removed genetically from their claimed Revolutionary ancestor than the Virginia Dowells are from the Maryland Dowells. This group of Dewells has not had a common paternal ancestor with any of the Dowells in this project for many thousands of years.

As this example makes clear, much of what we thought we had "learned" about family origins in the 20th century has had to be unlearned in the first decade of the 21st century as we add information provided by DNA testing. Isn't genealogy fun?

yDNA DESCENDANT TREE

A descendant tree has been developed that is based on testing at least two living descendants of each of three of the sons of Philip Dowell, Sr., the earliest Maryland Dowell who can be clearly documented. Philip died in 1733. By triangulating his descendants' test results, it has been established that many of the currently living descendants of the third son still have exactly the same 67 Y-chromosome marker values that Philip carried. Most of the descendants of the oldest son have mutated on only 1 or 2 markers over the ensuing generations. Descendants of the second son show mutations on as many as 4 markers of 67 tested over the three centuries since Philip Sr. lived. If nothing were known of the paper trail of these descendants of son number 2, the lab reports would show a 45 percent probability of common ancestor within eight generations. In this case the match would appear to be correct, even though the odds are slightly against it. It is good to keep in mind that probabilities are averages. Other information needs to be combined with DNA results to arrive at the best genealogical answers. No documented living descendants have yet been identified for son number 4, although he is known to have had children who survived him at his death in 1750. If we ever find any, we will strongly encourage them to join our DNA project.

Figure 2.7 shows some of the living descendants who trace their ancestry back to Philip Dowell, who died in 1733 in southern Maryland. The men at the bottom of each column have all taken yDNA tests. The number after their names shows how many markers they had tested by 2010.[16]

In Figure 2.8, the markers with mutations from the pattern for Philip Dowell, the founder, appear in shaded boxes. Also, note that both descendants of the first son share one mutation at marker 447. This would appear to indicate that the mutation on DSY 447 had occurred by the time "Maryland John" Dowell (1763–1812) contributed his DNA to his sons. Both of his living descendants in the project share that mutation. From this, we can hypothesize that any other descendants of Philip Dowell who show a value of 25 on DYS 447 might be expected to be descendants of "Maryland John" until other evidence disproves this possibility. However, only one of them shows a mutation on DYS 393, which indicates that this variation occurred sometime downstream from "Maryland John." Any Dowell who shows a value of 25 on DYS 447 and also a value of 14 on DYS 393 should be examined to consider the possibility that he descends from "Maryland John's" son William T. Dowell (1788–1855).

Figure 2.7
Living descendants of Philip Dowell (died 1733) of Maryland who had taken yDNA tests by 2010

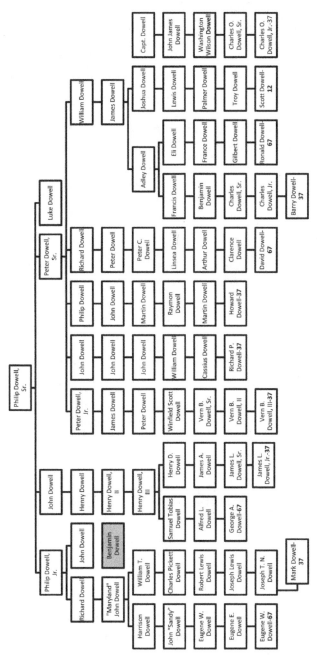

The only line of descendants who have remained in Maryland to this day was the one that experienced the most mutations. In fact, one of those tested for this line was born on property purchased by Philip Dowell, Sr., in 1722. Both descendants of the second

Figure 2.8
Markers deviating from the pattern of Philip Dowell, the founder, appear in shaded boxes

Maryland Dowell DNA	393	390	19	391	385a	385b	426	388	439	389i	392	458	389ii	459	455	454	447	437	448	449	464a	464b	464c	464d	460	H4	YCAIIa	YCAIIb	456	607	576	570	CDYa	CDYb	442	438
1st son	13	25	14	11	11	14	12	12	11	13	13	29	17	9	10	11	25	15	19	28	14	15	15	17	11	10	19	23	16	15	18	16	37	37	12	12
1st son	14	25	14	11	11	14	12	12	11	13	13	29	17	9	10	11	25	15	19	28	14	15	15	17	11	10	19	23	16	15	18	16	37	37	12	12
2nd son	13	25	14	11	11	14	12	12	11	13	13	29	17	9	10	11	26	15	19	28	14	15	15	17	11	10	19	23	16	15	19	16	37	38	12	12
2nd son	13	25	14	11	11	14	12	12	11	13	13	29	17	9	10	11	26	15	19	28	14	15	15	17	11	10	19	23	16	15	19	16	38	38	12	12
3rd son	13	25	14	11	11	14	12	12	11	13	13	29	17	9	10	11	26	15	19	28	14	15	15	17	11	10	19	23	16	15	18	16	37	37	12	12
3rd son	13	25	14	11	11	14	12	12	11	13	13	29	17	9	10	11	26	15	19	28	14	15	15	17	11	10	19	23	16	15	18	16	37	37	12	12
3rd son	13	25	14	11	11	14	12	12	11	13	13	29	17	9	10	11	26	15	19	28	14	15	15	17	11	10	19	23	16	15	18	16	37	37	12	12
3rd son	13	25	14	11	11	14	12	12	11	13	13	29	17	9	10	11	26	15	19	28	14	15	15	17	11	10	19	23	16	15	18	16	37	37	12	12
3rd son	13	25	14	11	11	14	12	12	11	13	13	29	17	9	10	11	26	15	19	28	14	15	15	17	11	10	19	23	16	15	18	16	37	37	12	12
3rd son	13	25	14	11	11	14	12	12	11	13	13	29	17	9	10	11	26	15	19	28	14	15	15	17	11	10	19	23	16	15	18	16	37	37	12	12
undocumented	13	25	14	11	11	14	12	12	11	13	13	29	17	9	10	11	26	15	19	28	14	15	15	17	11	10	19	23	16	15	18	16	37	37	12	12
undocumented	13	25	14	11	11	14	12	12	11	13	13	29	17	9	10	11	26	15	19	28	14	15	15	16	11	10	19	23	16	15	18	16	37	37	12	12
undocumented	14	26	14	11	11	14	12	12	11	13	13	29	17	9	10	11	26	15	19	28	14	15	15	17	9	10	19	23	16	15	18	16	37	37	12	12

son share mutations on markers 576 and CDYb. These mutations appear to have been acquired by the time Henry Dowell III (1827–1881) passed his Y-chromosome DNA to his sons. Each of Henry's descendants to be tested accumulated an additional mutation but at different locations on the chromosome. Again, this suggests that the additional mutations are more recent in origin. Such patterns can also be useful in placing those tested in the future who might not have established a paper trail.

We are still in the early days of learning why such mutations occur when they do. A variety of causes have been suggested, including illness, drug use, famines, wars, other environmental factors, and the age of the father at the time the child is conceived. So far, only this last factor has been found to have some scientific basis.[17]

None of the descendants of the third son who had been tested when Figure 2.7 was created in 2010 had mutations up to 67 markers. One line that was subsequently tested did have a couple of mutations that appear in recent generations.

The shared mutations appear to have occurred further back in time and were inherited by subsequent descendants. Mutations that were not shared with the nearest tested cousin appear to be more recent mutations that occurred after the lines of descent separated. Of the three undocumented family members, it is possible that one or more of them descends from the fourth son or that some of them descend from a so far unidentified brother or cousin of Philip Sr. They could also have descended from one of the first three sons. More work will have to be done before they can be placed with confidence.

RECONSTRUCTING THE yDNA OF AN ANCESTOR

By now you may be getting the idea that it is often possible to reconstruct the Y-chromosome signature of a long-dead ancestor. This can be done even without access to that ancestor's body or any other artifact carrying his DNA. This is best done within the context of a surname project such as the one discussed for the Dowells. The minimum requirement is that the ancestor had at least two sons who have unbroken male lines of descent to the present. In some cases three or more unbroken male lines of descent from that "founding" ancestor may be necessary to create confidence that the backward reconstruction is accurate.

In the simplest example, multiple sons of the same man are tested and their Y-chromosome results are identical. It can be hypothesized that the father contributed those alleles intact to each of his sons and thus shared with them his own DNA signature over those markers. In contrast, if two sons do not agree on every marker tested, additional sons (or male descendants of those sons) must be tested to try to determine what the pattern of the founder must have been.

The limit of this form of reconstruction is the paper trail from living men back to a common male ancestor. Generally this trail follows a common surname or at least connects surnames known to have a common origin. Sometimes DNA can add supporting evidence for relationships between men with different but similar surnames. On paper, a living male Groff and a living male Grove are ninth cousins twice removed. An exact 37 Y-STR match tends to confirm that they did, in fact, share a 17th-century ancestor. From documentary sources we know that ancestor was born in Baretswil, Zurich, Switzerland, and moved to Steinfurt, Baden (now Germany), to avoid persecution for his dissident Anabaptist religious beliefs. It is highly probable that this ancestor, born January 28, 1615/6, shared the exact 37 markers that were inherited by his two 21st-century descendants. Based on this match, TiP[18] calculates a 99 percent probability of a common ancestor within the number of generations these two men are shown to be separated based of traditional genealogical research.

RECONSTRUCTING THE FOUNDER'S 111-MARKER DNA SIGNATURE

Surname projects can apply these principles to reconstruct more complex descendant trees of family "founders." In the example given earlier in this chapter, the starting point was a man who had three known sons, all of whom have living male descendants of the same surname. In this case the "founder" was Philip Dowell, who appeared in the records of southern Maryland in the 1690s. By that time he was an established tobacco planter. His marriage in 1702 and death in 1733 are well documented, as are many other events in his adult life. However, no birth or christening records have yet been discovered. Although he is reported to have had four sons who passed his Y-chromosome DNA forward, living male Dowell descendants of only three of those sons have been identified and tested.

Five of those descendants have tested to 111 Y-STR markers (Figure 2.9). One was a descendant of Philip's first son, one was a descendant of Philip's second son, and three were descendants of Philip's third son. The results of these tests were that all five agreed on 101 of the 111 markers. In addition, four of the five (including descendants of at least two of Philip's sons) shared the same value on all 111 markers. In other words, none of the two mutations of the descendant of the eldest son, the six mutations of the second son, or the single mutations of two descendants of the third son occurred on the same marker. Philip can be assumed to have passed down the marker values that at least four of his five living descendants share because they were passed down through multiple lines that have no common ancestor more recent than him.

Figure 2.9
Triangulating 111 yDNA markers back to the founder of the Maryland Dowells

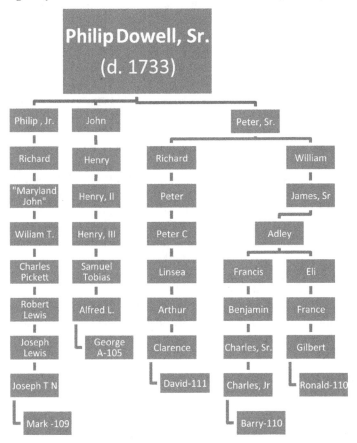

It is possible to establish the presumed DNA pattern of a founder with results from only two lines of descent. This only works, however, if those two sets of results from descendants agree completely. While this is often the case, such unanimity often does not extend over many generations. The reason the three sons are usually necessary to establish the DNA pattern of a founder is that often one or more mutations will occur as the DNA is passed to the next generation. This may be the result of a copying error that has occurred within the father as he ages. Molecular genealogist Angie Bush explains:

> DNA is not copied during fertilization and/or transmission of sperm to egg. DNA is copied as it goes through meiosis to produce gametes. In males, meiosis continues to occur throughout a man's life. As a man gets older, the copying of the DNA loses fidelity—this is known as the "paternal age effect." The best way to describe this might be that the "copy machine" that makes the sperm gets tired and worn out, and the older copies aren't as good as the first ones made by the copy machine.[19]

If that occurs, an older son may receive the old DNA value for that location and a younger may inherit a new value.

In the preceding example, a total of 111 markers were passed down through 41 intergenerational transfers from fathers to sons. Of the 4,551 separate data transfers, a total of 10 copying variations occurred, giving an error rate of 0.002197 or about two-tenths of 1 percent. In this very small sample, the mutation rate is similar to what has been found in much larger studies. These copying errors among the descendants of Philip Dowell were not randomly distributed, however (see Figure 2.9). In this sample, six of the ten errors occurred on the line of one descendant and took place in only seven generations downstream from the founder. One individual had two errors in nine generations; one had one error over nine generations; one had one error over eight generations; and one had no variation from the founder over eight generations on any of the 111 markers.

These results are a good example of the fickleness of probabilities. One of the lines—that of the descendant of the second son—accounted for more than half of the mutations identified. Descendants in a line like this one would have reason to doubt whether they were really related to their family. The descendant of the first son had a mutation on the very first marker tested and two mutations over the first 18 markers reported. If only his first 20 markers were considered, his connection to his legitimate family might be called into question. However, he had no additional mutations over the next 93 markers. The three descendants of the third son had a combined total of only two mutations. These two mutations did not show up until the 79th marker in one descendant and until the 100th marker in the other.

Making quick decisions about the validity of relationships when only a small number of markers have been tested can be dangerous. As a small reminder of this risk, note that one man in the Dowell example was only a 16/18 match with the DNA signature of the founder. He is shown as the descendant of the first son, Philip, Jr., in Figure 2.9. That result might make one suspect he was really not a member of this family. However, when the same descendant was later shown to be a 109/111 marker match, little doubt remained about who his seventh great-grandfather really was.

In Figure 2.9, the number after the names of those tested indicates how many markers, of 111 possible, were shared with the founder.

What good is this kind of information for a genealogist, you might ask? Within the context of this surname DNA study, it was hoped that identifying this Y-chromosome signature of Philip Dowell would facilitate matches with others outside the project that would help find his European origins. So far, that goal has not been achieved. None of the handful of Dowells from the United Kingdom and Australia has come close to matching any of the North American Dowells or one another.

"MARYLAND JOHN" DOWELL

Back in the 20th century, Clark Christoffersen and Amy (Dowell) Christoffersen, his mother, were trying to find the origins of their ancestor John Dowell, who passed away in 1812 in Albemarle County, Virginia. Family lore had said that his father's name was Richard Dowell. Unfortunately, both Richard and John have been popular given names among Dowell families. "After considerable research and a continuing process of elimination, it seemed that indeed Richard Dowell of Albemarle County (b. ca. 1745–d. 1836) was our 'most likely' progenitor."[20]

Their search for proof took them to the Virginia State Archives.

> "After obtaining permission to look, we searched through around two hundred and fifty boxes of sooty, un-indexed court records of all descriptions, none of it catalogued or organized. It was a fascinating, exciting and rewarding treasure hunt. . . . Disconcertingly, rather than confirming our much hoped for expectations," the case documents showed that the court order book's association of John as Richard's son had been a 1790's scribal mistake.[21]

Generally, multiple men of the same name in the same locality become like quicksand for successful genealogical searches. In this case, however, it was the proliferation of "John" Dowells in the county that gave Clark and Amy their first solid clue to their correct ancestral line. In reviewing microfilms of old tax records from around 1800, they noticed the name John Dowell appeared frequently. Apparently more than one man of this name lived contemporaneously in the county during the period covered by the records. The tax collector needed some memory cues to keep track of which John had paid his taxes each year, so he assigned nicknames to those with redundant names. We do not know if these names were inventions of the tax collector or if he merely recorded nicknames already in use in the vernacular. Whatever the case, we are very glad he wrote them down. The Christoffersens wrote, "And as it turned out, our John Dowell was described as of 'of Mero.' (Mero. Is an abbreviation for Meroland, a misspelling of Maryland)."[22]

Clark and Amy do descend from "Maryland John." Not only did this nickname assist them in separating his records from the other John Dowells in Albemarle County, but it also gave them an important clue to the origins of "Maryland John." However, at the time the Christoffersens made their discovery, it was mistakenly interpreted by some as confirmation that the Maryland Dowells and the Virginia Dowells were indeed closely related. It would take the Y-chromosome surname project to sort all this out.

We now have confirmed that "Maryland John" was, indeed, from Maryland; that by coincidence he did settle in the county that had the greatest concentration of Virginia Dowells; and that his descendants migrated farther west after only a generation or two with their Maryland Dowell DNA intact. Ironically, his father, who died in Maryland when John was still a teenager, was named Richard. We do not know whether these two groups of Dowells interacted with each other or even if they knew of each other's existence. They were concentrated in opposite corners of the county, a distance of some 20 to 30 miles.

One of Maryland John's descendants entered the project thinking he descended from an Albemarle County William Dowell who was not the son of Maryland John. William was also an overworked given name among the Dowells, in addition to John and Richard. The DNA test results clearly showed that this man was a descendant of Philip Dowell and not of the Virginia Dowells. Once he was attached to the correct William— William T. Dowell—the rest of his line fell into the correct alignment.

Another man joined the DNA project solidly convinced that he was a Virginia Group 1 Dowell. He had visited relatives and viewed graves of his assumed ancestors. His DNA results tell a different story. Apparently, this man's ancestors had not lived in Virginia at all until the 20th century and he was a Maryland Dowell. Now he has a whole different set of family reunions to attend. Other Dowells have also been connected with their ancestral line about which they had no previous clue.

INTERNATIONAL MAN OF MYSTERY

In the summer of 2009, the cover of *Ancestry* magazine asked, "What was our international man of mystery up to?" Featured in this issue was the story of a man who had abandoned two separate families in the United States in the early part of the 20th century. He then changed his name and moved to Australia, where he started another family. Long after he died in 1970, his Australian daughter, Lucinda Gray, set out to find out what she could about her father and his suspected earlier life under a different name and on a different continent. In the process, the daughter conducted extensive research and enlisted the help of many researchers. Many paths were followed to dead ends. Lucinda convinced her brother to take a Y-chromosome DNA test. No close matches appeared in the database for DNA Ancestry, the company that performed his test. Lucinda entered his results in Ysearch.org. As discussed earlier, this database allows Y-chromosome results to be compared when they have been conducted at different labs. In 2006, as I was researching my stepsons' Smithey line, I noticed that the DNA results posted by Lucinda were a fairly close match with those of my stepson Jason. I corresponded with Lucinda, and we tried to find any possible connection between the known Smithey ancestry and the life story of Lucinda's father as he had told it and she had been able to reconstruct it. No obvious points of contact emerged. The precision of the DNA match was hard to evaluate because the tests had been performed by different labs and not all of the markers tested by one lab had been tested by the other.

Lucinda did not tell the researcher working on her case about this DNA match until more than two years later, toward the end of 2008. Although the DNA match was not precise enough to make an exact identification, the surname with which it was associated caused Lucinda's researcher to revisit all the evidence she had collected and to determine that the man of mystery had been born with the name Smithers. Much more about what has been discovered about this man of mystery can be found in a 2009 *Ancestry* article.[23] The family maintains a website with many photographs, including two different copies of one picture: one copy was in the possession of the American family and one in the possession of the Australians.[24]

When the story broke, family members on both sides of the world reacted and their story was chronicled by Ruth Sheehan of the Raleigh (North Carolina) *News & Observer*:

> John Smithers of Raleigh had spent more than six decades looking for clues about the father who abandoned him, his sister and their mother when he was just a baby. The barrel-chested, brash-talking Smithers had something he wanted to give his old man: a fist in the nose.
>
> At 82, he had about given up on ever learning what happened to James William Smithers. He had long suspected his father got in trouble with the law and fled abroad. Decades ago, it was easy enough to disappear, and Smithers' father had seemingly vanished into thin air.
>
> On the other side of the world, Lucinda Gray had always wondered what her father's life was like before he moved mysteriously from the United States to Australia. She had spent years just trying to find out his real name.
>
> In mid-December, Smithers and Gray learned their elusive fathers were one and the same.
>
> After years of poring over records online and in person, across continents and oceans, it took only a DNA sample from a simple cheek scrape to bring the two branches of the family together.[25]

VARIANT SURNAMES BUT MATCHING yDNA

While precision and consistency are highly valued among serious genealogists, each of them can be taken to an extreme beyond which they begin to inhibit progress. Our language and our histories existed in oral form long before they were written—at least by the general population. In surname research, it is important to examine all possible variations of a given name. The Smothers Tribe surname project is a good example. There it has been discovered that most, but not all, Smither and Smithey men in the project are closely related, as is one Smiddy man. In fact, the two whose results best define the marker results of this cluster consists of one Smither and one Smithey. The men in this cluster can trace their ancestral lines back to Colonial Virginia or to nearby counties that bordered Virginia in Kentucky and North Carolina in the first decade of the 19th century. All the evidence collected so far points to a common ancestor in early Virginia for at least some of the lines, with the others joining them within a generation or two earlier on the other side of the Atlantic. However, where in the British Isles that might be depends on which tea leaves one is examining. If the match with the Smithers man of mystery is as close as it may be, perhaps these men also once had ancestors living in Sussex or Surrey in the south of England, as did our mystery man Smithers (also known as Gray). That is the area from which Smithers' ancestors immigrated to Canada before they ended up in Michigan.

SURNAME PROJECTS

To get the most benefit from your results, there really is strength in numbers. This can best be realized by joining both surname and geographic projects, some of which have been discussed in this chapter. This is clearly a case where the whole is greater than the sum of its parts.

FROM STRS TO SNPS

In early 2014, the focus of cutting-edge Y-DNA research was shifting from examination of short tandem repeats (STRs) to single nucleotide polymorphisms (SNPs). STRs are repeats (or, some say, stuttering) of the genetic code where the same sequence of letters is repeated over and over. Some of you will remember that such a sequence became the title of a 1997 science fiction movie (*Gattaca*). By counting the number of times such sequences of code are repeated in succession at a given location, we arrive at one of the STR values for that individual. By judiciously choosing the right mix of locations, we have been able to address many of our genealogical questions. STR values continue to be valuable.

SNPs are permanent changes in the genetic code where a G becomes an A, a C becomes a T, or vice versa. When these mutations occur, G always interchanges with A and T always interchanges with C. SNPs can be viewed as junctions where the DNA flow of one individual branches off from that of the rest of his or her tribe. This new SNP is then inherited by that person's subsequent descendants. Population geneticists and anthropologists use SNPs to trace our ancestors back through prehistory. At first Y-SNPs were thought to occur too infrequently to be useful to genealogists. However, as an increasing number of Y-SNPs are discovered, that is being called into question.

THE COMING SNP TSUNAMI[26]

The International Society of Genetic Genealogists (ISSOG) is a voluntary organization that does not charge dues. Since 2006, it has been responsible for maintaining the Y-DNA Haplogroup Tree for researchers and testing labs around the world. The number of Y-SNPs being discovered has been exploding since the end of 2010, and this accelerated discovery is just the beginning. The recent wave of newly discovered Y-SNPs has resulted from the Walk the Y,[27] Geno 2.0,[28] and 1,000 Genomes[29] projects as well as the normal discovery processes of investigation by academics and citizen scientists.

In November 2013, coordinator Alice Fairhurst of the ISOGG Tree Project reported the following startling growth statistics.[30] During the first decade of Y-DNA testing, fewer than 1,000 SNPs were reported and listed on the ISOGG tree. During the first nine months of 2013, more than 1,500 additional SNPs were added. From September 2013 to April 2014, about twice as many SNPs were added as had been added during the entire seven previous years, and the deluge is just beginning. Figure 2.10 documents this incredible growth curve of SNP discoveries.

Figure 2.10

Explosion in the number of SNPs listed in the ISOGG Y-DNA Tree

Time (End of Period)	Cumulative Number of SNPs in Tree
2006	436
2008	790
2010	935
2012	2,067
September 2013	3,610
April 2014	Almost 10,000

The tsunami has yet to come ashore. National Geographic's Geno 2.0 project has not yet published all its Y-SNPs. Treasure troves of additional Y-SNPs from Chromo2 Full-Genomes, and FTDNA's Big Y tests loom just over the horizon in early 2014 as this manuscript is being bundled off to the publisher. These have the potential to identify and place tens of thousands of heretofore unknown Y-SNPs. Many of these will be leaves toward the ends of branches on the Y-DNA haplogroup tree. We hope many of them will be recent enough to connect with the family trees that have been documented by genealogists. Will these new terminal SNPs become unique enough to become the equivalent of earlier coats of arms as identifiers of specific family lines? We may find out before the end of this decade.

These Y-SNP tests are a supplement to—not a replacement for—the Y-STR tests that have been our staple until now. As FTDNA explains in its FAQs, the BIG Y is not for beginners who are new to DNA testing. At this time, it is too complex and is aimed at bridging the gap between genealogy and deeper ancestry. Those seeking matches for genealogical purposes still should begin with Y-STR testing:

> The BIG Y product is a direct paternal lineage test. We have designed it to explore deep ancestral links on our common paternal tree. It tests both thousands of known branch markers and millions of places where there may be new branch markers. We intend it for expert users with an interest in advancing science.
>
> It may also be of great interest to genealogy researchers of a specific lineage. It is not however a test for matching you to one or more men with the same surname in the way of our Y-DNA37 and other tests.[31]

SUMMARY

Y-chromosome testing, in its current state of refinement, can be very useful in verifying paternal lines, in exposing incorrect lines, and in giving clues to guide further research. It is more conclusive in disproving relationships than it is in establishing a definitive line of descent. It will not by itself prove descent from a specific distant patriarch. The DAR is correct in not accepting such test results, in the absence of supporting evidence, as proof of eligibility for membership. At the same time, society would be wise to recognize that a wide variance in Y-chromosome test results disproves previously accepted qualifying lines such as the Dewell line mentioned earlier in this chapter. In other words, many closely related men over several generations can share the same or very similar Y-chromosome DNA. However, men whose Y-chromosome DNA test results vary widely cannot be close biological relatives along their paternal line.

A Y-chromosome test can also predict a male's paternal haplogroup—but that is a discussion for another chapter, as is how both women and men can select closely related surrogates to assist them in using Y-chromosome testing to explore some of the paternal lines in their pedigree charts. For an impartial and up-to-the-minute comparison of various commercial Y-DNA testing options, check the ISOGG website: http://www.isogg.org/wiki/Y-DNA_testing_comparison_chart.

NOTES

1. John H. Works, Jr., "A Primer on Jefferson DNA," http://www.pbs.org/wgbh/pages/frontline/shows/jefferson/true/primer.html, accessed August 31, 2012.

2. Angie Bush, communication with the author, March 1, 2014.

3. This British tradition is covered in detail in George Redmonds, Turi King, and David Hey, *Surnames, DNA, and Family History* (New York: Oxford University Press, 2011).

4. Sorenson Molecular Genealogy Foundation, "History," http://www.smgf.org/history.jspx, accessed December 28, 2013.

5. Sorenson Molecular Genealogy Foundation, "About the Sorenson Molecular Genealogy Foundation," http://www.smgf.org/pages/overview.jspx, accessed December 28, 2013.

6. Tim Janzen, "An Important Update on SMGF from Dr. Tim Janzen," *Your Genetic Genealogist* (June 30, 2012), http://www.yourgeneticgenealogist.com/2012/06/important-update-on-smgf-from-dr-tim.html, accessed December 28, 2013.

7. There are other STRs in the autosomal DNA, but they do not have DYS#s in their names.

8. Georgia K. Bopp, "Non-Paternal Event (NPE)," *Kinney, McKinney & Variations, Lenhart & Variations, DNA Project Notes,* December 15, 2006, http://freepages.genealogy.roots web.ancestry.com/~gkbopp/DNA/falsepaternal.htm, accessed December 5, 2013.

9. Triangulation refers to the process of comparing the marker results of two (or more) living men known to have a common ancestor. It is assumed that if the DNA of each of the living descendants matches that of the other, it would agree with the DNA of the closest male ancestor whom the men share in common. This old surveyors' technique was applied to genetic genealogy a decade ago by William Hurst in a post on Genealogy-DNA-L: http://archiver.rootsweb.ancestry.com/th/read/genealogy-dna/2004-12/1103242991, accessed October 5, 2013.

10. The probability was calculated by TiP, a copyrighted process of Gene by Gene, Ltd.

11. Ibid.

12. Chris Pomery, "The Advantages of a Dual DNA/Documentary Approach to Reconstruct the Family Trees of a Surname," *Journal of Genetic Genealogy* 5, no. 2 (2009): 86, http://www.jogg.info/52/files/Pomery.htm, accessed July 18, 2011.

13. All testing in this project was completed through Family Tree DNA.

14. All statistical probabilities quoted herein were calculated by the FTDNA TiP. These results are based on the mutation rate study presented during the 1st International Conference on Genetic Genealogy, on October 30, 2004. The probabilities take into consideration the mutation rates for each individual marker being compared.

15. DAR Genealogical Research System, "Descendants Database Search," http://services.dar.org/public/dar_research/search_descendants/default.cfm, accessed January 25, 2013.

16. Since 2010, many of those represented in this chart have been tested for additional markers and other Dowell men have been tested. These additional results do not change the basic findings, but they do make it difficult to represent them clearly on a two-dimensional sheet of paper.

17. Hayley Dunning, "Father's Age Affects Mutation Rate: The Number of New Gene Mutations in Children Rises Dramatically with the Age of Their Father at Conception," *The Scientist* (August 22, 2012), http://www.the-scientist.com/?articles.view/articleNo/32515/title/Father%27s%20Age%20Affects%20Mutation%20Rate, accessed October 18, 2012.

18. The probability was calculated by TiP, a copyrighted process of Genealogy by Genetics, Ltd.

19. Angie Bush, communication with the author, March 1, 2014.

20. Clark D. Christoffersen and Amy Beth Dowell Christoffersen, *In Search of John Dowell: A Dowell Family History, 1620–2004* (Spokane, WA): Gregath, 2003), 273.

21. Christoffersen and Christoffersen, *In Search of John Dowell,* 274.

22. Christoffersen and Christoffersen, *In Search of John Dowell,* 274.

23. Colleen Fitzpatrick, "One Man, Two Names, Three Families and Much Intrigue," *Ancestry* (July–August 2009): 22–31.

24. "James William Smithers Finally Found after More Than 60 Years … ," http://jakehendersongray.com/, accessed January 25, 2013.

25. Ruth Sheehan, "DNA Reveals Story of Dad's Disappearance," *Raleigh News & Observer* (March 23, 2009). http://www.newsobserver.com/2009/03/23/43491/dna-reveals-story-of-dads -disappearance.html , accessed July 30, 2014.

26. David R. Dowell, "ISOGG Group Gears up for SNP Tsunami," *Dr. D Digs up Ancestors,* http://blog.ddowell.com/2013/11/isogg-group-gears-up-for-snp-tsunami.html, accessed November 13, 2013. I appear to be the first person to publish the phrase "SNP tsunami," although others may have used it previously in informal communication.

27. "Walk through the Y," *ISOGG Wiki*, http://www.isogg.org/wiki/Walk_Through_the_Y, accessed December 30, 2013.

28. *National Geographic,* "Geno 2.0: The Greatest Journey Ever Told. Your Story. Our Story. The Human Story," https://genographic.nationalgeographic.com/, accessed December 30, 2013.

29. "About the 1000 Genomes Project," *1000 Genomes: A Deep Catalog of Human Genetic Variation,* http://www.1000genomes.org/about, accessed December 30, 2013.

30. Alice Fairhurst, Report at ISOGG meeting during 9th International Conference on Genetic Genealogy, November 20, 2013, and Alice Fairhurst, email to the author, April 6, 2014.

31. FTDNA, "What Is the BIG Y product?", http://www.familytreedna.com/faq/answers .aspx?id=27#2205, accessed November 25, 2013.

3

Who Is the Mother? "Umbilical" mtDNA

For the most current generation, it is easier to determine with certainty who the mother is than it is to know who the father is. The biological mother is required to be present at a birth event—except in the rare case of the use of a surrogate. The mother is generally the one from whom information is gathered to generate the birth certificate. It is possible that she is uncertain who the father is or wants to disguise his identity—much to the dismay of adoptees and others seeking to find the identity of their birth parents. Since in most Western cultures a clear majority of women change their names when they marry and often when they divorce, this advantage is soon lost to those researching the maternal lines of preceding generations.

There's good news for women or for anyone who wishes to research his or her female lines but felt left out of the discussion of Y-chromosome DNA in Chapter 2. "Did you know that there is also a DNA test that traces your direct *maternal* line back in time? It is called a mitochondrial DNA (mtDNA) test."[1] For genetic genealogists, the mtDNA test results provide a mirror image of the results provided by a Y-chromosome test. Additional good news about mtDNA testing is that individuals of both genders can use it to explore their direct maternal line ancestors—as long as no male interrupts that female line of ascent. Other features of mtDNA should also be kept in mind. Specifically, mtDNA is "abundant and durable but less variable"[2] than the DNA found in the nucleus of human cells. The "more abundant" feature means that it is more likely to survive in human remains in adverse conditions than other kinds of DNA because of the thousands of copies that often exist in a single human cell. The "less variable" feature is a double-edged sword. While mtDNA is not as precise in differentiating between closely related individuals, it is good for tracing the deep ancestry of groups of people.

The possibilities of mtDNA for use by genealogists have been touted since the early 1990s. In a seminal article that appeared in the prestigious *American Ancestors*, Thomas Roderick, Mary-Claire King, and Robert Charles Anderson wrote:

> Genetic technology has advanced at an explosive rate in the last few years. A major technical advance of interest to genealogists is the capability now to identify, without doubt, first-degree relatives (parents, full siblings or children) of any given individual.

This capability extends as well to more distant relatives, but with lower assurance the farther the genealogical relationship.

Studies of mtDNA could also be of significant genealogical value.

Most DNA is inherited equally from each parent. The mtDNA, however, is different in that it is inherited principally from the mother; that is, each of us, male and female, inherits rather specific mtDNA only through his or her matrilineal (dubbed "umbilical" by some genealogists) line. It is *ONLY ONE* ancestral line. Thus, for example, in the fourth generation back I inherit my mtDNA from only [one] individual, my mother's mother's mother's mother, and she from *her* mother, and so on. Furthermore, I and all my mother's children, as well as anyone who has a matrilineal line that intersects my matrilineal line, will share the same mtDNA. To be sure, as we go much further back we would encounter mutations or changes that took place in the mtDNA. But in general, our mtDNA describes our matrilineal heritage for many generations. Comparisons of mtDNA among us can potentially verify the matrilineal lines we have constructed through genealogical research and point to other, more distant relationships.[3]

This article ignited interest in genetic genealogy in a number of the current leaders of this field, including Ann Turner. With Meghan Smolenyak, Turner later authored one of the first comprehensive books in this field.[4] Ann wrote at Tom Roderick's death in 2013:

It was Tom's prescient article "Mitochondrial DNA: A Genealogical and Genetic Study" in the 1992 NEHGS journal that inspired me to get involved in DNA testing. It excited me so much that I visited him at his home in Bar Harbor before DTC testing was even available and saw some of the original traces of lab work validating the concept with some long New England mtDNA lines.[5]

Roderick was credited with coining the term "genomic" and with cofounding the first online genetic database.[6] He is also credited with coining the visually descriptive phrase "umbilical line DNA" to describe the matrilineal transmission of mitochondrial DNA.

Mitochondrial tests explore the maternal ancestral line. This line is traced by the dark boxes in Figure 3.1. In reality, such tests probe deep ancestry along that line far beyond the generations shown in this simplified illustration.

"Mothers pass mtDNA to their children, both sons and daughters, but only females pass it on. Your mtDNA was inherited from your mother and from her mother and from her mother,"[7] back any number of generations in which you inherited DNA from a woman to which her descendants were attached by an umbilical cord. "No matter how far back in time you go, you only have one direct maternal ancestor in each generation and she is the one responsible for passing you your mtDNA. Your mtDNA has followed this matrilineal path down through the generations for many thousands of years—intact and virtually unchanged."[8] Ultimately, this genetic material came from "Mitochondrial Eve" (mtEve). This mtEve is not to be confused with the Biblical Eve, whom many consider to be the mother of all humans. Instead, mtEve is the earliest woman who had two (or more) daughters, each of whom has had an unbroken female line of descendants down to the present day. In other words, mtEve is the earliest woman who has passed mitochondrial DNA down to currently living humans along at least two separate lines of descent.

Although mtDNA changes very slowly over thousands of generations, many mutations have accumulated over the eons. The slow rate of change allows us to follow it back into time. At the same time, the mutations, as they accumulate, allow us to distinguish between different paths of human migration.

Figure 3.1
Mitochondrial tests explore the maternal ancestral line

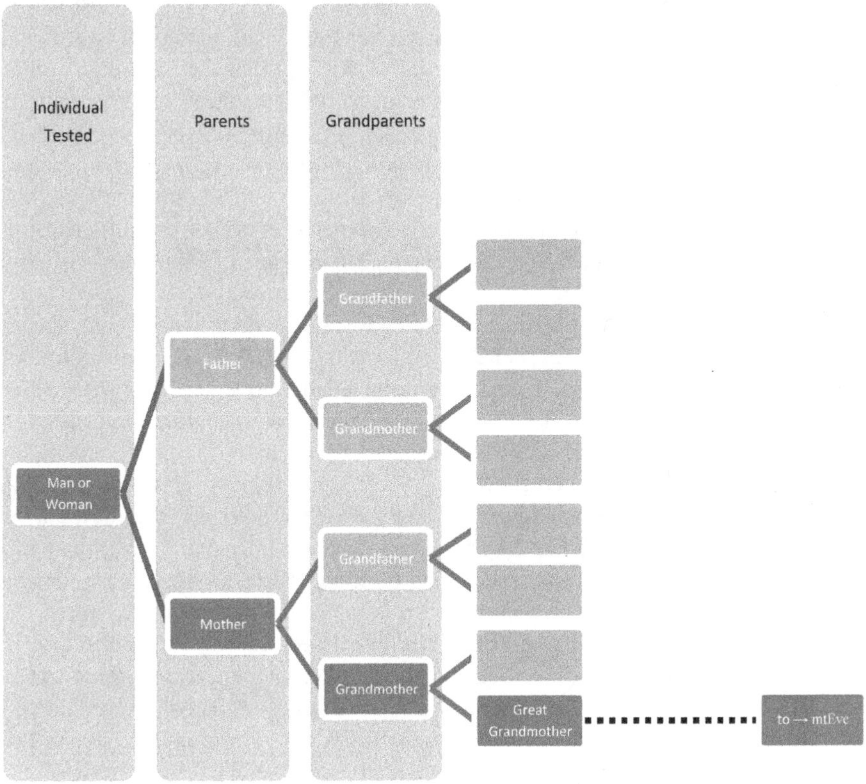

Mitochondria are tiny compared with other areas of the cell that contain DNA. With *only* 16,569 locations (compared to the total of more than 3 billion locations in the nucleus of each cell), mtDNA is small enough for genetic genealogy labs to record and report information from each of those locations. However, even 16,569 bits of information is too much for most human minds to organize and interpret. As a result, it is much easier to communicate only the locations at which two individuals differ.

What is being examined is which of the four possible chemical bases in DNA is found at each location. Those bases are adenine (A), cytosine (C), guanine (G), and thymine (T). Fortunately, it is not necessary to understand the chemistry of the cell to interpret the results of an mtDNA test for genealogical purposes. Instead, it is necessary to know only at which locations SNPs have occurred—single places where the mtDNA sequence differs between individuals across this portion of their genomes.

Direct-to-consumer tests for mtDNA have been marketed at three levels. The first to be offered to genetic genealogists was for hypervariable region 1 (HVR1). HVR1 is part of the so-called control region of mtDNA. In the past, it was thought to be an area of mtDNA in which more than an average amount of variation occurred. As HVR1 was part of the control region as opposed to the coding region, researchers thought that results found here could be freely shared between individuals without disclosing any sensitive heritable information.

The area examined in an HVR1 test is locations 16024 to 16569 on the mtDNA circle. Although this area covered 546 bases and in mtDNA terms was considered a hot spot for variation, it turned out to be less than satisfactory for the average genealogist. If two individuals shared an exact match over HVR1, all that could be concluded was that there was a "coin flip" (50 percent) probability that they shared a common female ancestor in the last 52 generations or roughly 1,300 years.[9] While the failure to match might eliminate many potential relationships, most genealogists are focused on the last three or four centuries and are looking for more certainty. In other words, the lack of a match at this level would mean it was highly unlikely that two individuals were related along their umbilical lines; an exact match, however, simply means they were related sometime—perhaps in the deep dark past. Half of such matches are predicted to have occurred prior to 700 AD.

As the cost of testing dropped, an expanded level of testing was offered. This test covered hypervariable region 2 (HVR2)—the locations 00001 to 00576 in addition the locations covered by HVR1. Since mtDNA is arranged in a circle, this segment is actually adjacent to HVR1 and is also part of the control region. For genealogical purposes, this level of testing represented an improvement but was still nothing to get excited about. An exact match at the combined HVR1 and HVR2 level merely raised the probability that two individuals with an exact match actually shared a maternal ancestor to that of a coin flip sometime during the last 28 generation or about 700 years.[10] How many of the lines on your family tree are you researching back around 1300 AD? At that time surnames were just beginning to be used by a few families in widely scattered areas.

Around 2009, the price of a test covering the entire mtDNA dropped to the point where it was almost affordable to serious genealogists. This test covered all 16,569 bases included in both the control region and the coding region. The increased precision of an exact match moved the coin-flip probability of a common ancestor to those who clearly would have lived within the genealogical age. That is, there was now a 50 percent chance that individuals who found an exact match shared a common female ancestor within the last five generations.[11] Many of us have been able to extend our genealogical research back that far at least on some of our lines. It is important to remember that there is also a 50 percent chance that the common ancestress lived further back in time than that—in some cases much further back. Four hundred years would not be an unreasonable expectation for some such matches.

REVISED CAMBRIDGE REFERENCE SEQUENCE

The first full mtDNA sequence to be documented was completed by a group of geneticists at the University of Cambridge, England, and the results were published in *Nature* in 1981.[12] Actually, this sequence was not quite complete:

> The mtDNA sequence was first published by a research group at Cambridge University, and the Cambridge Reference Sequence (CRS) was the standard against which other mtDNA sequences were compared for almost two decades, Interestingly, however, the authors could not obtain the complete sequence of the mtDNA from their human subject, and therefore filled in some gaps in the sequence with sequences from bovine [cow] mtDNA and the mtDNA from a cancer cell line called HELA that is commonly used as a research tool.[13]

Not only is it interesting that cows are similar enough to help us map human mitochondria, but it is also worth noting the contribution of HeLa cells.[14] HeLa refers to "a cell type in an immortal cell line used in scientific research. It is the oldest and most commonly used human cell line. The line was derived from cervical cancer cells taken on February 8, 1951, from Henrietta Lacks, a patient who eventually died of her cancer on October 4, 1951."[15] The life of Henrietta Lacks, an impoverished African-American woman, and the contribution she unwittingly gave to all of us through her cells are chronicled in the excellent book *The Immortal Life of Henrietta Lacks* by Rebecca Skloot. This 2010 book raises many important questions about race, ethics, and medical research.

By the end of the 1990s, researchers were able to decode the last few locations of human mtDNA and revise the original Cambridge Reference Set:

> In spite of the fact that only 11 nucleotides needed revising, the revision presented a problem, because the C nucleotide at position 3106 in the original CRS is missing from the revised CRS [rCRS]. In order to avoid having to change the numbering of all nucleotides after 3106, the rCRS includes an "N" in position 3107 as a place holder. The rCRS is currently used as the reference sequence.[16]

With only minor modifications in 1999, the rCRS remained the standard against which all other mtDNA results were reported until recently. In other words, all subsequent mtDNA results were reported by listing only those locations where the results differed from those discovered in the first British effort three decades ago. For those of modern European maternal descent, this simplified the reporting because they had relatively fewer differences since they were being measured against another modern European.

This artificial starting point gave us a way to communicate and compare mtDNA results among individuals. Although the extent to which one's results varied from this standard could be measured, the derived distance had no intrinsic value. It was sort of like Greenwich Mean Time, a standard point that is recognized around the world, but in the last analysis is not better or worse than Eastern Standard Time. Use of rCRS created the mistaken impression that all mtDNA evolved from the reference individual in Cambridge. As we saw with the fable of the Blind Men and the Elephant in Chapter 1, your perspective defines what you see.

RECONSTRUCTED SAPIENS REFERENCE SEQUENCE

This Eurocentric view of mtDNA was challenged in April 2012, when geneticist Doran Behar and eight colleagues published in *The American Journal of Human Genetics* an article entitled "A 'Copernican' Reassessment of the Human Mitochondrial DNA Tree from Its Root."[17] For those of you who may have forgotten exactly why Copernicus is important, he proposed the idea that we revolve around the Sun instead of the other way around. In so doing, Copernicus drastically changed our perspective of how we on Earth relate to the rest of our universe. Behar et al. proposed that we measure our mtDNA from the perspective of how it deviates from mtEve rather than how it deviates from an anonymous person in Cambridge. Later that year, Family Tree DNA—the primary lab doing mitochondrial DNA sequencing for genealogical purposes—implemented the Reconstructed Sapiens Reference Sequence (RSRS).

Figure 3.2

Sample of full mitochondrial test results from the rCRS perspective

Your Results

| RSRS Values | rCRS Values |

HVR1 DIFFERENCES FROM rCRS				HVR2 DIFFERENCES FROM rCRS				CODING REGION DIFFERENCES FROM rCRS				
16519C				152C	263G	309.1C	315.1C	750G	1438G	2259T	4745G	4769G
								7337A	8660G	13326C	13680T	14872T
								15326G				

Revised Cambridge Reference Sequence

HVR1 REFERENCE SEQUENCE			HVR2 REFERENCE SEQUENCE			CR REFERENCE SEQUENCE		
Show All Positions			Show All Positions			Show All Positions		
Position	CRS	Your Result	Position	CRS	Your Result	Position	CRS	Your Result
16519	T	C	152	T	C	750	A	G
			263	A	G	1438	A	G
			309.1		C	2259	C	T
			315.1		C	4745	A	G
						4769	A	G
						7337	G	A
						8660	A	G
						13326	T	C
						13680	C	T
						14872	C	T
						15326	A	G

TEST RESULTS REPORTS

If you were to take an mtDNA test, what would your report of results look like? Figures 3.2 and 3.3 offer two examples.

Figure 3.2 shows the full mitochondrial results reports from the revised Cambridge Reference Segment perspective. The result reported in the left column of Figure 3.2 (16519C) indicates that the only difference between the tested individual and the Cambridge reference person in HVR1 (between locations 16024 and 16569) is found at location 16519. At location 16519, the mtDNA of the Cambridge individual showed a "G" (guanine) while our tested individual showed a "C" (cytosine). Although this difference helps us separate lines of descent, it may have no known significance for health or physical characteristics of the two individuals. The HVR2 region and the coding region results can be interpreted in a similar manner.

Figure 3.3 shows full mitochondrial test results reported from the revised Reconstructed Sapiens Reference Sequence perspective. Which report would you prefer to receive? Can you spot the difference between these two sets of results? They are actually for the same individual and both report exactly same information, albeit

Figure 3.3

Sample of full mitochondrial test results from the RSRS perspective

Your Results

| RSRS Values | rCRS Values |

| Extra Mutations | 309.1C | 315.1C | 522.1A | 522.2C |
| Missing Mutations | C152T | | | |

HVR1 DIFFERENCES FROM RSRS					HVR2 DIFFERENCES FROM RSRS					CODING REGION DIFFERENCES FROM RSRS				
A16129G	T16187C	C16189T	T16223C	G16230A	G73A	C146T	C195T	A247G	309.1C	A769G	A825t	A1018G	C2259T	G2706A
T16278C	C16311T				315.1C	522.1A	522.2C			A2758G	C2885T	T3594C	G4104A	T4312C
										A4745G	T7028C	G7146A	T7256C	G7337A
										A7521G	T8468C	T8655C	G8701A	C9540T
										G10398A	T10664C	A10688G	C10810T	C10873T
										C10915T	A11719G	A11914G	T12705C	G13105A
										G13276A	T13326C	T13506C	T13650C	C13680T
										T14766C	C14872T			

from two different perspectives. HVR1 in Figure 3.3 shows seven mutations from mtEve. One of them is not location 16519. Therefore, the apparent "mutation" reported at that location in Figure 3.2 was a mutation from mtEve in the Cambridge individual and not in the individual whose test results appear here. The later individual shares the same value as mtEve at that location.

Genetic genealogist CeCe Moore describes the implications of such results:

> Mitochondrial DNA testing is a great tool for discovering more about the females in your family tree. As all dedicated genealogists have undoubtedly experienced, female ancestors are frustratingly difficult to trace because of the fact that they traditionally change their surnames with marriage [and sometimes with divorce.] So, for many of us, anything that may shed light on our female ancestors' origins is welcome. The value of mtDNA testing is based in the simple fact that if two people share an exact mtDNA sequence, then those two people descend from a common female ancestor somewhere in the past.
>
> For each individual, your mtDNA will *only* reflect that of your direct maternal line. This applies to both males and females. However, following the same rule that applied to Y-DNA, any female ancestor of interest can be followed down the tree to the present day in an attempt to find an appropriate direct maternal descendant for mtDNA testing.[18]

Mitochondrial DNA was the first type to be tested to trace ancestry, but it was not the first to gain the attention of genealogists. As mentioned in Chapter 1, it generally is the most durable form of DNA, so it is the most likely to survive the passage of time and the ravages of the elements. If DNA is recovered from fossils, it is most likely to be mitochondrial DNA. This factor also makes mtDNA the best candidate for helping to identify the human remains from old plane crashes, unknown soldiers, and other victims of disasters recovered long after these tragic events occur. Although it traces the female line, it can be used to find information on males up that female line, as the following story illustrates.

In a tale that gives new meaning to the old phrase "skeletons in the family tree," researchers in England recently conducted an exhaustive search to find living relatives of Richard III, the last Plantagenet King of England. In 1485, Richard became the last British monarch to die in battle. Archeologists recently unearthed bones in Leicester suspected to be those of the king who died in the battle of Bosworth Field. "The crooked spine of a long-dead warrior, complete with an arrow in its back and a gash across its skull, was found on the site of the Grey Friars church, where King Richard III is thought to have been buried."[19] This dramatic description turned out to be partly incorrect. The arrow turned out to be a nail "that was apparently mixed in with the remains" according to Richard Buckley, the project's lead archaeologist.[20] Mitochondrial DNA was the genetic vehicle chosen to investigate the possibility.

On February 4, 2013, the University of Leicester held a news conference to announce that these remains had been proved beyond a reasonable doubt to be those of Richard III.[21] Alan Boyle, science editor at NBC News, wrote:

> The team's genetic analysis reinforced the link to Richard III: DNA was extracted from bone samples and compared with modern-day mitochondrial DNA from two direct descendants of Richard III's family, including an anonymous donor as well as Michael Ibsen, a Canadian-born cabinetmaker who is a 17th-generation descendant of Richard III's eldest sister, Anne of York.

Genetic matches based on mitochondrial DNA aren't as clear-cut as, say, a paternity test—but a mismatch would have ruled out any family connection. Similar techniques were used to identify the remains of Czar Nicholas II and other members of Russia's royal family, who were killed in 1918 during the Russian Revolution.[22]

IN THE RECENT ANALYSIS

The DNA being used in the test is mitochondrial DNA (mtDNA), which is the DNA outside of the cell nucleus. Unlike nuclear DNA (nDNA), which is inherited half from the mother and half from the father, mtDNA is inherited from the mother alone. Geneticists prefer to use mtDNA for testing partly because they have identified the markers they need to use; and partly because a female line doesn't have the risk of being corrupted by illegitimacy.[23]

YOUR MATCH LIST

If you take an mtDNA test, the most interesting thing in your results—at least at first—will not be which value is reported for you at a specific location. You will want to know who in the database may be closely related to you. After all, you are a genealogist. With mtDNA, your matching can be either deluge or famine—but ideally will be somewhere in between. My wife of the last 30 years, who tested her complete mtDNA in 2009, has yet to find a match at any level of testing. We suspect this is because her maternal line comes from an area of Eastern Europe where few people have tested. Or perhaps she really is an alien being, as she sometimes claims.

My ex-wife, and the mother of my children, is more clearly of this planet. She currently has 37 exact matches on her full mtDNA and a total of 87 matches within two mutations. The most matches I have encountered are those of my daughter-in-law's father. He has more than 300 matches over his entire mtDNA, almost 100 of which are exact. His match total is inflated by his Ashkenazi ancestry: that group is currently present in DNA databases in numbers much greater than their proportion of the total world population would suggest. While his maternal ancestresses may have been fruitful and/or over-sampled, that has not been the case for his paternal line. He has only one exact match at 25 markers and no matches within two mutations at 37 markers. Whether this apparent imbalance in matches between his maternal and paternal lines will ultimately be explained by genetic defects, epidemics, or other external events remains to be seen. Both his father's and mother's lines appear to have migrated to New York toward the end of the 19th century from the same general area near Kiev, but details of their lives in Europe have yet to be discovered.

My own mtDNA test was initially discouraging. After testing in 2009, I had no matches on my complete mitochondria for almost three years. On full mitochondria tests in the early days, FTDNA only reported exact matches. This was frustrating because near matches as well as exact matches were reported at the lesser levels of HVR1 and HVR2, even though such matches appeared to be genealogically useless. My first match turned out to be worth the wait. I first blogged about this match in March 2012.[24]

I finally have an exact, full segment, mitochondrial DNA match—at all 16,569 locations along my mitochondria! That may not be a big deal to some people. To me it was. I have been waiting about three years for it.

I was not alone. No exact matches had shown up for my wife or for my father-in-law, either. Then I was notified that I had not one match, but two. As it turns out, the two were essentially one. I matched a mother and her son who would be expected to carry identical mitochondrial DNA. However, that match is turning out to be significant genealogically speaking.

I have long known from documentary research that I have a line of ascent that goes back to Anabaptist nonconformists who were in Switzerland prior to 1600. Relying on the research of others, I have been able to identify four eighth-great-grandparents who were part of this sect. My mitochondrial match was from Switzerland. My earliest confirmed direct maternal line ancestor was my great-great-grand mother Mary Ann (Shover) Grove, who married into the line that descended from the Anabaptist nonconformists.

From my newly discovered distant cousin I got this email, which is shared with her permission:

If I had to make a wild guess, I'd say that descendants of our common ancestress had moved to the States due to religious persecution. My great-grandmother was a very religious person. She and her husband must have originally come from Baptist families who had publicly abdicated their faith generations before but secretly kept on with their beliefs. They kept a 400-year-old "Froschauer" Bible in their trunk. This Bible was forbidden in Berne after the Reformation for a long time and only Baptists kept on using it. This particular Bible had all the verses of the New Testament marked in red ink, which suggests that it was used for more than just Bible reading at home. I got this piece of information from a book about the farm on which my great-grandmother and her husband had lived on, written by Hand Schmocker.

Then two days later:

I have just hung up the phone with the administration of citizenship in the town, Langnau im Emmenthal, where my great-grandmother, Rosalie Gerber, was born in 1871. (Her husband's ancestors I can easily trace back to 1652: the actual family tree already exists, published in the book I mentioned yesterday.) I made an appointment at the citizen office for April 30th to study all the possible registers there, starting with my great-grandma and then going back. I assume that this will consequently lead me to other towns where more looking up will be necessary. I talked to my aunt on the phone and she will accompany me since she knows much better how to read the so-called "Kurrentschrift" in which all the registers around here before the 20th century were written in.

Thanks for your fascinating information about your ancestors. The Anabaptist movement started in Zurich beginning of the 16th century. But the first Anabaptists also already appeared in 1525 in Bern. The ancestors you described all carry surnames which are not indigenous to the area my great-grandmother is from. However, my great-grandmother carries a typically Bernese surname indigenous to the Emmenthal. So there is a story there to trace.

People moved in all directions once the persecution started. Quite a few moved from the Zurich area towards the mountains (Emmenthal, Berner Oberland); hence so many still live around here where we live (our neighbors in front are "neo-Anabaptists"). But there is no knowing in which direction (to or from the mountains) our common ancestress or her offspring have moved.

I will try to find out more. Well, wish me luck on my endeavor.

This is turning out to be a DNA match that was well worth the wait. For now, my hypothesis is that the ancestors of *both* Mary Ann (Shover) Grove and her husband Samuel Grove, Jr., were part of the Anabaptist clan in Switzerland 400 years ago. I know that Samuel's ancestors had fled/migrated to the Baden area (now Germany) by 300 years ago and were in Lancaster County, Pennsylvania, a generation later. Samuel's line subsequently spent two generations in the Shenandoah Valley of Virginia before moving to Licking County, Ohio, where Samuel was born in 1818 and where he married Mary Ann in 1840.

Mary was apparently born in Pennsylvania about 1822. I do not know if her family knew Samuel's family before each migrated to Licking County. However, it is beginning to look like this colony of Anabaptists may have been very close knit. Was it chance or clan taboo that led them to marry someone within the group? In any case, the gene pool may not have been as wide as one might think.

Since 2012, my mtDNA Swiss cousin Gabriela has been able to extend her maternal line one generation earlier but there is still no indication that she has found our "missing link" ancestress. In addition, I now have one more exact full mtDNA match—Gabriela's daughter, who of course matched her mother.

FROM WHENCE CAME MARJORY?

Geneticist and genetic genealogist Angie Bush reminds us how to begin a research process:

> The first step of any research process (scientific or genealogical) is to define the question being asked. Once that question is well defined and understood, then that directs the researcher in the direction to go next and the resources to exploit in order to find the answer. Maximizing the use of DNA testing for genealogical purposes requires that the question of kinship and identity being asked be well defined prior to testing.[25]

Defining the research question will not only help you decide if DNA testing has the potential to help answer your question, but should also guide the choice of the appropriate test and suggest who in the family should be tested. While I was researching the origins of one of my sixth great-grandmothers, I developed such a question that my father's DNA—not mine—would have helped to answer. That grandmother happened to be on my father's mother's umbilical line. My problem was that my father had passed away untested just a few years before.

Not to worry. One of my father's sisters has a living daughter who also connects directly—without male interruption—into that same umbilical line back to our shared sixth great-grandmother. My question, when I asked my cousin Ruth to swab her cheeks, was whether my sixth great-grandmother Marjory Oins/Owens was Swedish or Welsh. Marjory first showed up in the written record on January 8, 1736. On that day she married, as his third wife, Henry Stedham at Holy Trinity "Old Swedes" Church in what is now Wilmington, Delaware. Henry's grandfather had emigrated from Sweden to the North American colony of New Sweden in the middle of the 17th century. By the time of the wedding, the area formerly known as New Sweden had passed through the hands of the Dutch and was then controlled by the British. The community in which this wedding took place was still a pretty tight ethnically Swedish community. Note the name of the church where the wedding took place. Even so, Oins or Owens sounded Welsh to me. In fact, I have Welsh Owens ancestors who were living at the time, no more than 100 miles away as the proverbial crow flies, in Maryland.

So, I wondered, was Marjory Swedish or Welsh? As it turned out, she was neither—at least on her maternal side.

As soon as Ruth's sample was processed by FTDNA, she had three exact matches. One of the matches said he had been able to trace his maternal line only back to South Dakota, and the family tradition was that they were from Scandinavia. Two matches still live in Finland. A couple of months later, at the end of 2010, a fourth match, CeCe Moore, was added. Yes, *Your Genetic Genealogist* has traced her maternal line back to Margaretta Mattsdr Martenson Kauppi Storstaka, who was born July 21, 1712, in Härmä, Lansi-Suomen, Laani, Finland. Margaretta was a contemporary of my Margery.

So it appears that Marjory was neither Welsh nor Swedish. As far as I am concerned, Ruth's match with CeCe and the other three people convince me that Marjory was ethnically a Finn—at least on her maternal line. Since I found these matches, I have discovered that Finland was part of Sweden at the time New Sweden was settled. In addition, many historians believe that more than half of the colonists who came to New Sweden were ethnically Finns. At one time there was a settlement in New Sweden, now long extinct, that was named Finland.[26]

HETEROPLASMY

It turns out that CeCe and Ruth do not exactly share the same values on all 16,659 locations along their mitochondria. Soon after CeCe was listed as an exact match on Ruth's page, CeCe disappeared from that list. However, Ruth still appeared on CeCe's page. Both CeCe and her mother have a *heteroplasmy*. The National Institutes of Health (NIH) defines that as follows: "The situation in which, within a single cell, there is a mixture of mitochondria (energy-producing cytoplasmic organelles), some containing mutant DNA and some containing normal DNA."[27] Apparently the mtDNA of CeCe is in the process of mutating at one location, the only one that is not an exact match with my first cousin Ruth and also with my paternal grandmother.

In the early days of full mtDNA testing, the protocol for reporting such matches or nonmatches had not been settled. At the time, only exact mtDNA matches were reported to customers. Now those that are as many as three steps removed are reported on results pages and CeCe is shown to be a single step removed. Ruth currently has 48 matches within three steps, most of whom have been able to identify a European place of origin or still live in Finland (see Figure 3.4).

As a result of this new focus on early 18th-century Finland, Ruth and I have not yet been able to extend this root of our tree. We may never be able to do so, but at least we know where *not* to look. By comparison, another of CeCe's matches, who is still living in Finland, has helped her to extend her known family tree back four more generations.

In spite of a research trip to the Delaware State Archives in Dover and the Delaware Historical Society Research Library in Wilmington, I have been unable to learn anything about the life of Marjory prior to the time she married Henrick Stedham in 1736. I don't know whether Oins or Owens was her maiden name or the surname of a previous husband, although I suspect the former. Ruth's matches have stimulated me to learn a lot about the arrival of Finns in New Sweden and their culture, including their introduction, in the 17th century, of log cabin construction into the Delaware River Valley.

Ruth's matches convince me that my eighth great-grandmother and possibly my seventh great-grandmother down my paternal grandmother's umbilical line were not only ethnically Finns but possibly lived in Finland.

Figure 3.4
Location map of Cousin Ruth's full mtDNA matches

mtDNA AND atDNA TOGETHER

In most cases, mtDNA is not useful in pinpointing exact relationships between two individuals. Its slow mutation rate makes it a blunt research tool for time dating. While it weathers well over time, it is not able to differentiate among closely related individuals. However, when used in combination with autosomal results, mtDNA can be very useful in sorting out relationships in the last few generations.

I reported on one such application in a September 2013 blog post.[28] Two women whom I called Allyson and Bertha had taken both mtDNA and autosomal DNA (atDNA) tests, hoping to sort out their relationship. In particular, they wished to determine the birth mother of Bertha's mother, who had been adopted by Allyson's great-aunt in the 1920s. Family lore had assumed that it was an "in the family" adoption and that the biological mother was either Allyson's mother as the result of a teenage fling before she met Allyson's father or Allyson's grandmother in an "illicit liaison" with Allyson's grandfather sometime after they had been divorced. The atDNA test results pointed to a first or second cousin relationship between Allyson and Bertha. The amount of atDNA shared suggested that their relationship was most likely that of first cousins once removed. The mtDNA test supported the assumption that Bertha's mother had been adopted within the family and that Allyson and Bertha had a common female ancestress sometime in the recent past. These results taken together did not support either of the two scenarios that had circulated quietly through the family. However, they did support the possibility that Bertha's biological grandmother was also Allyson's grandmother if her former husband was not involved. In this case Bertha and Allyson would be half first cousins and would share the same amount of atDNA as would first cousins once removed. In resolving this family mystery, mtDNA played a minor but important role; the leading role was played by atDNA. Chapter 4 provides much more information about this kind of DNA in general and this match in particular.

AncestryDNA, 23andMe, and FTDNA all provide competitive atDNA tests in the U.S. marketplace. However, FTDNA is the only viable full-service mtDNA provider in North America. The company provides the largest database of tested individuals against which you can compare your results. Additional options exist in other parts of the world. Check the ISOGG website for the latest comparative information.[29]

Additional tools for analyzing your mtDNA results are linked to the ISOGG site.[30] These resources include PhyloTree, a widely used phylogenetic tree of global human mitochondrial DNA variation. This authoritative tool is used by both research scientists and experienced genetic genealogists. It is maintained by Mannis van Oven, who is a molecular biologist at the University Medical Center in Rotterdam, the Netherlands.[31]

SUMMARY

Mitochondrial DNA is inherited along the umbilical line from mothers by all their offspring of either gender. Men inherit it but do not pass it on. Because of the number of copies contained in each cell, mtDNA often survives in human remains long after other kinds of DNA have faded away. It can be extremely useful in tracing female lines deep into the past, but its slow mutation rate usually requires us combine its use with other tools to sort out close relationships.

NOTES

1. CeCe Moore, "DNA Testing for Genealogy: Getting Started, Part Two" posted by *Geni,* July 25, 2012.

2. Ron C. Michaelis, Robert G. Flanders, and Paula H. Wulff, *A Litigator's Guide to DNA: From the Laboratory to the Courtroom* (Amsterdam: Academic Press, 2008), 171.

3. Thomas H. Roderick, Mary-Claire King, and Robert Charles Anderson, "Mitochondrial DNA: A Genetic and Genealogical Study," *American Ancestors* (November 27, 1992), http://www.americanancestors.org/mitochondrial-DNA-a-genetic-and-genealogical-study/, accessed March 23, 2014.

4. Megan Smolenyak and Ann Turner, *Trace Your Roots with DNA: Using Genetic Tests to Explore Your Family Tree* (city?: Rodale, 2004).

5. Ann Turner, "Tom Roderick," post to the ISOGG listserv, September 13, 2013.

6. "Thomas H. Roderick Ph.D.," *The Bangor Daily News* (September 9, 2013), http://obituaries.bangordailynews.com/obituaries/bdnmaine/obituary.aspx?n=thomas-h-roderick&pid=166834450&fhid=6169#fbLoggedOut, accessed September 13, 2013 .

7. Moore, "DNA Testing for Genealogy."

8. Moore, "DNA Testing for Genealogy."

9. Family Tree DNA, "How Many Generations back Does Mitochondrial DNA (mtDNA) Testing Trace?", http://www.familytreedna.com/faq/answers.aspx?id=10#486, accessed January 29, 2013.

10. Family Tree DNA, "How Many Generations."

11. Family Tree DNA, "How Many Generations."

12. S. Anderson et al., "Sequence and Organization of the Human Mitochondrial Genome," *Nature* 290, no. 5806 (1981): 457–465.

13. Michaelis et al., *A Litigator's Guide to DNA*. Those of you wishing to see the detailed changes between CRS and rCRS are directed to Mitomap, "Revised Cambridge Reference Sequence (rCRS) of the Human Mitochondrial DNA,"http://www.mitomap.org/bin/view.pl/ MITOMAP/HumanMitoSeq, accessed March 8, 2013.

14. The original HeLa cells were contributed in 1950 by Henrietta Lacks, a poor African American woman with cervical cancer. The intriguing story of her cells and of their contributions to our lives was chronicled by Rebecca Skloot in *The Immortal Life of Henrietta Lacks* (New York, NY: Broadway Paperbacks, 2010). Skloot's fascinating account raises many issues about ethics in medical research, race, and poverty. The paperback edition is preferred because of the supplementary material added at the end.

15. "HeLa," *Wikipedia*, accessed March 17, 2014.

16. Michaelis et al., *A Litigator's Guide to DNA,* 171–172.

17. Doran M. Behar et al., "A 'Copernican' Reassessment of the Human Mitochondrial DNA Tree from Its Root," *American Journal of Human Genetics* 90, no. 4 (2012): 675–684.

18. Moore, "DNA Testing for Genealogy."

19. Martin Wainright, "Richard III: Could the Skeleton under the Car Park Be the King's?" *The Guardian* (September 12, 2012), http://www.guardian.co.uk/science/2012/sep/12/richard -skeleton-king-remains-bosworth , accessed January 31, 2013.

20. NBC, "Verdict Issued on Skeleton Found under Parking Lot: It's King Richard III," http:// cosmiclog.nbcnews.com/_news/2013/02/04/16832540-verdict-issued-on-skeleton-found-under -parking-lot-its-king-richard-iii?lite, accessed February 4, 2013.

21. BBC, "Richard III Dig: DNA Confirms Bones Are King's," http://www.bbc.co.uk/news/ uk-england-leicestershire-21063882 , accessed March 4, 2013.

22. Alan Boyle, "Verdict Issued on Skeleton Found under Parking Lot: It's King Richard III," *NBC News,* http://cosmiclog.nbcnews.com/_news/2013/02/04/16832540-verdict-issued-on -skeleton-found-under-parking-lot-its-king-richard-iii?lite, accessed February 4, 2013.

23. Joanna Moorhead, "Genetic Testing: To Catch a King: Michael Ibsen Is a Carpenter But He's Descended from Plantagenet Kings. A New Test, Using His DNA, May Help to Identify the Bones of Richard III—and Revolutionise the Study of Family History." *The Guardian* (December 7, 2012), http://www.guardian.co.uk/lifeandstyle/2012/dec/08/genetic-test-dna -richard-3-skeleton, accessed January 10, 2013.

24. David R. Dowell, http://blog.ddowell.com/2012/03/another-ana-baptist-ancestress.html.

25. Angie Bush, communication with the author, February 25, 2014.

26. Thomas Campanius Holm, A short description of the Province of New Sweden now called by the English, Pennsylvania, In America, translated by Peter S. Du Ponceau, M'Carty and Davis, *Memoirs of the Historical Society of Pennsylvania* 3, no. 1 (1834): 81.

27. National Institutes of Health, "Genetics Home Reference: Your Guide to Understanding Genetic Conditions," http://ghr.nlm.nih.gov/glossary=heteroplasmy, accessed September 20, 2013.

28. David R. Dowell, "Solving a Mystery with Women's DNA," *Dr. D Digs up Ancestors* (September 1, 2013).

29. ISOGG, "mtDNA Testing Comparison Chart," http://isogg.org/wiki/MtDNA_testing _comparison_chart, accessed December 2, 2013.

30. ISOGG, "MtDNA Tools," http://www.isogg.org/wiki/MtDNA_tools, accessed October 22, 2013.

31. M. van Oven and M. Kayser, "Updated Comprehensive Phylogenetic Tree of Global Human Mitochondrial DNA Variation," *Human Mutation* 30, no. 2 (2009): E386–E394, http://www.phylotree.org. doi:10.1002/humu.20921, PhyloTree.org-mtDNA tree Build 16 (February 19, 2014), accessed March 20, 2014.

4

Who Is Closely Related? atDNA

"SEX doesn't matter anymore."

With that provocative slogan, FTDNA announced its "Family Finder" test in 2010. In so doing, it joined 23andMe in the autosomal DNA (atDNA) market. Two months earlier, in February of that year, 23andMe had taken its comparable "Relative Finder" test from beta status to the general market. Two years later, Ancestry.com would enter this segment of the market. As attention getting as the "Sex" catchphrase may have been, FTDNA's other tagline—"Find family across all of your lines"—more accurately described these new genetic tests.

Now atDNA testing is where much of the action is—particularly for newcomers to genetic genealogy. It is the most competitive segment in the marketplace, with three major players slugging it out for market share in the United States and a few others making offerings in other parts of the world. As a result of this competition, the price point has dropped below the psychologically significant point of $100 and is often discounted further during sales. At this price it is possible to get results that allow you to become a viable player in genetic genealogy. It is also more practical to test multiple close family members. This is very beneficial with atDNA to an extent that is not nearly as productive when testing yDNA and mtDNA.

WHY TAKE AN atDNA TEST?

Irish genetic genealogist Maurice Gleeson identifies the following uses for information from atDNA tests:

1. To confirm known or suspected relationships
2. To connect with distant cousins
3. To use what they know about their family trees to break through brick walls in ours
4. To determine your ethnic makeup[1]

Historically, most genealogists have rated uses 1 and 3 as being of greater value than uses 2 and 4. As Australian genealogist Rhonda Lucas puts it, "We are growing our DNA presence in an attempt to prove/disprove our lineages and we are hoping it will take us back further."[2] However, each of us will have our own priorities, and these priorities may shift from time to time.

AUTOSOMAL DNA (atDNA) TESTING

Although yDNA and mtDNA potentially are very powerful tools in breaking through barriers that documents research is unable to resolve, often these tools cannot read the information in the part of our genomes that may contain the answers we are seeking. Taken together, they address only a very small part of one's total family tree. As researchers succeed in discovering additional generations of their ancestors, the fraction of the next generation that these tools can help with discovering plummets precipitously. While yDNA and mtDNA taken together can give a male information about both his parents, they return information on only half of his grandparents, one-fourth of his great-grandparents, and one-eighth of his great-great grandparents. Recall from the discussion in earlier chapters that DNA samples provided by women yield information on only half that many ancestors—those on the maternal side of the woman's family tree.

To help find information on all the grayed-out ancestors in the middle branches of a person's family tree (Figure 4.1), testing of atDNA was introduced to the public early in 2010. As shown in Figure 4.1, autosomal DNA helps find matches from descendants of the ancestors in the gray boxes that cannot be discovered by Y-chromosome or mitochondrial testing. It potentially provides information on every ancestor within the last few generations who contributed DNA to an individual.

HOW DO WE INHERIT atDNA?

Let's review some basic principles about how our identity is passed down to us. If you have studied a little genetics or carefully read Chapter 1, you may remember that each person has 23 pairs of chromosomes within each cell. Twenty-two of those pairs—the autosomes—give us most of our unique characteristics. They are inherited differently than the 23rd pair—the sex chromosomes. The somewhat more multifarious inheritance pattern of xDNA will be discussed in a subsequent chapter.

Y chromosomes are passed directly from fathers to their male off-spring *without mixing with the genes of the mother*. Likewise, mitochondrial DNA is passed directly from the mother to all her offspring *without mixing with the genes of the father*. In this direct transmission, both can pass down unaltered through many generations. That is the norm. However, the rare copying errors in these intergenerational transfers act as somewhat of a molecular clock to give an estimate of how recent in time a shared direct paternal or maternal line ancestor may have lived. These copying errors or mutations also allow us to trace different branches of the human family and place them in a phylogenic tree.

INHERITANCE PROCESS FOR AUTOSOMES

The transfer of DNA by the 22 pairs of autosomal chromosomes (sometimes called autosomes) from one generation to the next is quite a different process. The autosomes are numbered 1 through 22. We all receive a chromosome 1 from our father and a

Figure 4.1
Autosomal DNA tests trace all your family lines

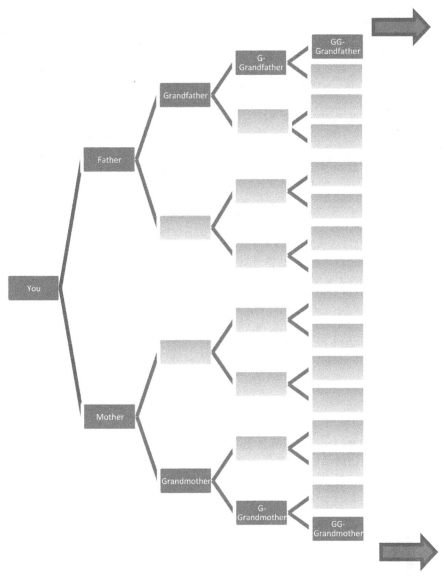

chromosome 1 from our mother, except in rare genetic abnormalities. This process continues all the way through chromosome 22. Prior to passing one of these chromosomes on to the next generation, a process called recombination shuffles the genetic material of these two original chromosomes, and the offspring receives one of these "shuffled" chromosomes.

Note that half of the total atDNA from the parents is transmitted to their child. But which half is it? Each of the parents will contribute half, but their halves will not necessarily be composed of exactly half of what each of the parents inherited from their parents, the grandparents of the new child. The pieces of atDNA that each parent, in his or her 50 percent contribution, passes down to a child are selected in what appears to be a random process from the atDNA that each previously have inherited from his or her own father and mother. It is unlikely that the new child will receive exactly the same amount of atDNA from each grandparent.

If this couple has another baby, the random process is repeated. Again each parent contributes to the process. However, after this process is completed, the half of the DNA that each parent passes on to the second child may be quite different from what the first child received. Hence, we have the difference between siblings. This is basis for why autosomal inheritance is sometimes so frustrating to interpret. All kinds of oversimplifications have been developed to help us understand how this inheritance pattern plays out in offspring. I am sometimes asked if DNA can "skip a generation." The simple answer is that, while DNA cannot skip a generation, how it is manifested in traits can. For example, a recessive red hair gene can pass unnoticed until at random it is matched up with another red hair gene from the other parent.

This random selection and passing down goes on generation after generation. During this process, some segments of DNA remain intact while others are chopped up to form multiple shorter sequences. Autosomal DNA testing, for genealogical purposes, looks for intact segments of DNA that are identical between individuals. If these segments are of sufficient length, they are likely to have originated from the same shared ancestor. The more of these segments that are shared by two individuals, the closer their biological relationship is thought to be.

Figure 4.2 provides a simple example of how atDNA might be passed from your great-grandparents down to you. Note that by the time your parents inherited their atDNA, they shared much more with some of your great-grandparents than they did with others. By the time the randomly selected atDNA reached you, some of the segments inherited by your parents had been passed down to you almost intact as they inherited them, but some were beginning to become almost too short to notice.

Figure 4.2
Chart illustrating the random nature of atDNA inheritance (Courtesy of Angie Bush.)

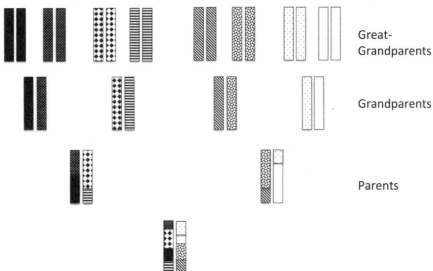

Figure 4.2 illustrates the random nature of how atDNA might be inherited from great-grandparents. A much more complete discussion of recombination and inheritance of atDNA can be found in Graham Coop's blog post at http://gcbias.org/2013/12/02/how-many-genomic-blocks-do-you-share-with-a-cousin/.[3]

Genetic genealogists will be interested in determining how recently the common ancestor lived who passed this segment down to both of you. Although we are not privy to the exact algorithms that the genetic genealogy testing companies use to predict relationships between two people, those algorithms use a combination of segment length and total amount of DNA shared to provide a "guess" as to how closely you might be related to another person. In the case of autosomal DNA, recombination also serves as a "molecular clock" being used to determine how long ago a common ancestor may have lived.

One thing is clear: the success of this kind of testing relies very heavily on access to comprehensive family trees that have been thoroughly researched from traditional documentary sources. No longer can one assume that a match lies on the paternal line, as is the case in yDNA. Likewise, one cannot assume that a match lies on the umbilical cord line, as is the case in mtDNA testing. Instead, the match could be on either of those lines *or on any line in between*. To take full advantage of atDNA matches, one needs a robust pedigree chart that is well researched along all ancestral lines. Once a shared segment is discovered, it is now time to start the tedious work of locating shared surnames and locations that may help you to identify where and when those people lived.

Such information can be compiled and documented only through traditional genealogical research. In some ways, autosomal results seem to say, "There is a common ancestor approximately x generations back on one of your lines. Now go find that person." Most of us will have a hard time taking full advantage of the power we are being offered by atDNA testing. Rather than being the "magic bullet" that by itself reveals our family trees, such testing reemphasizes the need for traditional genealogy research. Experienced genetic genealogist Larry Vick stated it this way: "Whether we use Relative Finder, Family Finder, or both, most of us need to do a much better job on our pedigrees to get more value from either test. I plan to spend a week at the Family History Library this summer beefing up my pedigree."[4] In my own experience, many of my autosomal matches are adoptees or others who know little about their family histories. For them and for those of us who discover genetic matches with them, learning how we might be related can be a very frustrating process.

What is the probability that you will share enough DNA with a relative that it will be detected by autosomal testing? If you are related within five generations (third or more recent cousins), then atDNA is very likely to detect your relationship. Testing will also detect many of your fourth cousins and some of your fifth and even more distant cousins. Combining the experience with autosomal testing by both FTDNA and 23andMe, the chances of finding a match with a particular cousin—that is, detecting a common segment of DNA between two specific related individuals—within the last few generations are shown in Figure 4.3.[5, 6]

Thus, if someone is a fourth cousin or closer, there is a good chance they will be detected in autosomal testing. Detection of matches further back is a longshot but is still possible.

Herein lays a paradox. Although you will have a small and decreasing probability of matching *a specific* fifth or more distant cousin, most of the matches for you that are identified by the lab will fall into these categories. This result occurs because the number of cousins you potentially could match increases geometrically as we go back through the generations. You have only 8 sets of great-great-grandparents who could

Figure 4.3
Probability of matching a specific related individual in an autosomal database

Relationship	Probability of Detecting Common Segment
Second cousins or closer	> 99%
Third cousin	90%
Fourth cousin	45–50%
Fifth cousin	10–15%
Sixth cousin and more distant	< 5%

have produced descendants who could be your third cousins, but you have 64 sets of great-great-great-great-great-grandparents who could have produced descendants who are your sixth cousins.

A major problem is that even those of us with well-researched pedigree charts rarely have been able to identify all 32 of our great-great-great grandparents. They would be the ancestors we have in common with a potential fourth cousin. In addition, our potential fourth cousins (with whom we have just been told we share an identifiable autosomal segment) would need to have documented all of their 32 great-great-great grandparents as well. After researching my family tree for almost five decades and working at it diligently and continuously for the last two decades, I know the identities of only 23 of my own 32 great-great-great grandparents. This issue explains why it is difficult to identify which ancestors are the ones with whom we share our atDNA matches. Beyond the fifth generation, many matches will occur *and* common ancestors will show up in the pedigree charts of those of us who have well-researched trees. However, it becomes increasingly difficult, to the point of becoming nearly impossible, to determine whether that matching segment of atDNA was actually inherited from a couple whom we find both in our own pedigree chart and in the pedigree chart of the person with whom we share that match. The fact that a couple appears in both your pedigree chart and that of the distant cousin with whom you share a common segment of atDNA does not necessarily mean that the pair is the ancestral couple from whom the two of you inherited that atDNA. You could have other ancestral connections—even though they may be yet undiscovered. It also is possible that the two of you could share identifiable segments inherited from more than one ancestral pair.

MAJOR atDNA TESTING COMPANIES FOR GENEALOGISTS

In the United States, three companies currently dominate the testing of atDNA to further family history research: AncestryDNA, Family Tree DNA, and 23andMe. The most important thing each offers to command our attention and our dollars is its database of tested individuals. Each database provides a pond in which we as genetic genealogists can go fishing. The size of these pools and the methods of fishing allowed and enabled are the key reasons why you would choose one company over another. The best scenario would be to fish in all three simultaneously. The remainder of this chapter highlights some of the features that are useful when you begin to explore your matches. Not all companies provide all these tools—but each offers some of them. I have attempted to use different tools as I moved from company to company, thereby giving you the widest range of ideas in the shortest amount of time. You will soon learn which of these tools are useful when you investigate your matches at whatever company site you are using.

FISHING IN THE ANCESTRYDNA POND

AncestryDNA came to the atDNA testing field only in 2012. This recent arrival may account for what some seasoned genetic genealogists believe to be a rather underdeveloped set of tools to enable users to analyze and interpret the information encoded into their atDNA. But even if this criticism is at least partially true, is this sufficient to cause us to shun the company's product? Blaine Bettinger blogged this warning:

> A word of advice: beware anyone who tells you to avoid AncestryDNA. Many genetic genealogists, myself included, have had incredible success using AncestryDNA's autosomal DNA test. Personally, several of my own major DNA discoveries have occurred though the service. Unfortunately, it has become popular among some genetic genealogists to deride AncestryDNA's autosomal DNA test, and some recommend avoiding the service altogether.[7]

To this post, British genetic genealogist Debbie Kennett responded:

> What you say is fair enough for Americans but the biggest problem with AncestryDNA as far as I'm concerned is that their test is only sold in America. It's fine for Americans who want to match other Americans. It's of no use to anyone living in any other country in the world because they cannot buy the test even if they wanted to do so.[8]

This shortcoming affects all of us, as Maurice Gleeson commented from Ireland: "So it is an excellent resource for anyone with Colonial US ancestors but comparatively less useful for those trying to connect with ancestors who came to the US in the last 100–200 years, and less useful still for anyone without any US connections."[9] Those of us who are looking for the DNA match that will take us back across the pond may have to wait a while longer.

My own experience with fishing in the AncestryDNA pond is improving. Since I have been writing this book, my own personal genealogy research time has been somewhat curtailed. As a result, I have not had the opportunity to pursue all possible leads with the vigor they may deserve. I am constantly bumping into matches who remind me that many who have undergone such testing do not have the slightest idea how to exploit their results to benefit their research. For this discussion I am lumping test takers into three categories:

1. Those who are serious genetic genealogists
2. Those who are serious genealogists but have little or no knowledge of how to apply test results to assist their family history research
3. Those who took the test but know little about ether genealogy or genetic testing

It is great to get a match with someone who is in category 1. The exchange of information can be very efficient and mutually beneficial. Those in category 3 are difficult to help unless you are willing to do some serious mentoring. They are likely to find useful information for themselves and to provide useful information to their matches only if the matches are very close—second cousins or closer—and their matches know some family history. The ones in category 2 are probably the most numerous—at least among those with whom you may share a "leaf match."

Shaky Leafs

The greatest friend to customers of Ancestry's DNA test is the shaky leaf. Generally at Ancestry.com, "The Shaky Leafs are Ancestry Hints designed to aid you in your family history research. They provide other records in our database that could possibly match the individual in your tree."[10] The leaf image appears to be intended to show areas where our trees can "grow" by adding new information from Ancestry. In the case of a leaf appearing in conjunction with a DNA match, it means that Ancestry's search algorithms have found a person in the pedigree tree uploaded by your match that appears to be the same person who is also in the pedigree tree you uploaded. This is an ancestor who is suggested to be one shared by the two of you. Finding a new shaky leaf connection with a person with whom you share a DNA match is indeed a happy time. As far as I am concerned, this is the best feature of AncestryDNA. Of course, you need to verify that this hint is correct, but it removes much of the tedium from the matching process.

Serious genetic genealogists will also want to verify the location and length of that match on the genomes and to explore whether others may match at that same location. To date, AncestryDNA has not provided customers with the kind of tools that allow these kinds of inquiries to be conducted on its site. To pursue this level of analysis, one needs to download the raw data from the Ancestry site and upload it to FTDNA (for a small fee) and/or upload it to a third-party site like www.gedmatch.com (free). These uploads, particularly to FTDNA, offer the additional advantage of allowing one to fish in another of the big three ponds, opening up the possibility of finding even more cousins who have tested there but not at AncestryDNA. FTDNA currently is the only company that accepts the transfer of raw data from other labs[11] without requiring that the user submit a new sample to be tested.

By early 2014, I had received 21 leafy matches from AncestryDNA. My wife had received none. Her lack of matches results in part because five of her eight great-grandparents lived in Prussia/Germany until late in the 19th century. Ancestry has, at least until now, tested only in the United States, which limits the number of distant cousins she could potentially have in the company's database. FTDNA does test overseas. 23andMe does as well but its shipping costs have limited participation outside North America.

Figure 4.4 shows some of my atDNA matches at Ancestry.com, all of whom shared a "shaky leaf." Names of test takers and others who may be managing their results have been removed to maintain privacy.

Figure 4.5 is a blow-up from Figure 4.4 that shows a new match distant cousin at AncestryDNA with locked pedigree chart. Note the padlock and leaf in the upper-right corner of this figure. Although the computer algorithms have compared our charts and discovered a likely shared ancestor, only my match can see which ancestor meets that criterion unless I am able to get an invitation to that chart, which often is problematic.

However, the match predictions generally are correct as is the case shown in Figure 4.6.

Figure 4.4
Screen shot of atDNA matches at Ancestry.com

Of my leafy matches, I have so far been able to identify 19 common sets of ancestors I share with my cousins with whom I have a matching DNA segment (Figure 4.7). Of course, by definition I need to have a robust tree that contains all of these multiple great-grandparents; if I did not, the cousins would not have been auto-identified by Ancestry as leafy matches. If my tree or the trees of the cousins had not contained some of these ancestors, we could not have been leafy matches.

Figure 4.5
Blow-up of an atDNA match from Ancestry.com

In the previous paragraph I said I have been able "to identify 19 common sets of ancestors I share with my cousins with whom I have a matching DNA segment." This is not quite the same as saying that the 19 common sets of ancestors so identified were the sources of our matching segments of DNA. We do share DNA and we do have matching sets of ancestors. However, we also could have other as-yet-unidentified shared ancestors; perhaps they are the sources of our shared DNA. That is probably not the case with these matches but it is a possibility we must keep in mind. We will come back to this situation later in this chapter. Currently AncestryDNA does not provide analysis tools that allow users to examine the specific DNA segments in more detail.

Figure 4.6
A "review match" of a predicted shaky leaf match

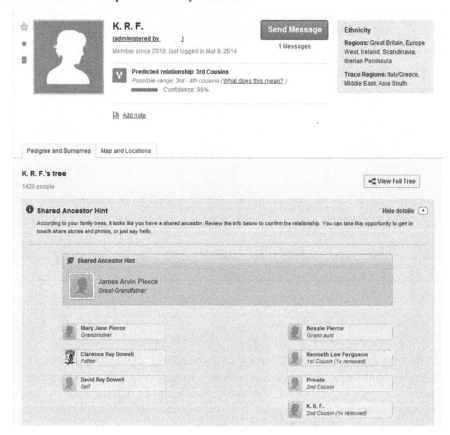

Figure 4.7

Predicted and actual relationships of cousins who have DNA matches *and* automated pedigree chart matches at AncestryDNA

AncestryDNA's Predicted Relationship to Me	Actual Relationship to Me
Third cousin (1)	Second cousin once removed (1)
Fourth cousins (3)	Third cousin once removed (1)
	Fifth cousin (2)
distant cousins (15)	Third cousin twice removed (1)
	Fifth cousin (2)
	Fifth cousin once removed (1)
	Sixth cousin (4)
	Sixth cousin once removed (1)
	Seventh cousin (2)
	Seventh cousin once removed (4)

My pedigree chart begins to break down significantly with my third great-grandparents. I think I know the identities of all my second great-grandparents. However, I know the full identity of only about two-thirds of all my third great-grandparents. For Ancestry's system to help you identify new cousins who can help you expand your family tree, *both* of you already must have robust pedigree charts and *both* of you must be willing to share. One of my pet peeves is people who post locked trees on Ancestry.com and then take a DNA test. Getting access to the locked trees of matches does not work well both because of inattention of the tree owners and the efficiency of Ancestry's method of extending invitations the owner wishes to offer. If you really feel the need to have a private tree *and* take Ancestry's atDNA test, consider posting another tree that is open and contains only the names, dates, and locations of known direct-line ancestors. If you come across a private tree in Ancestry's system, send a message requesting that the tree be shared with you. Although you may not get a response, many people are willing to share if they are asked.

I have been able to identify three additional cousins apart from the leafy match system discussed previously. One was identified as a fourth cousin, but his mother is actually my third cousin. She and I had previously exchanged family information and pictures. As a result, this third cousin once removed remembered my name and contacted me after he had been notified of our DNA match. His match confirmed a relationship that we had previously documented by other means but did not really add new information.

Nevertheless, two matches have added important new chapters to my family history. One of these could have been a leafy match except that variant spellings of the given name of our common third great-grandmother sufficiently disguised her from Ancestry's computers. I had to find this match the old-fashioned way—by going through the cousin's tree name by name. Combined with considerable traditional research, this discovery blew away one of my long-standing brick walls. I was immediately able to extend this branch of my tree by a couple of generations. Subsequent research has now appeared to extend this line back several additional generations to 16th-century Scotland. I am now in the process of trying to verify all of these extensions. So far, they look very good. As I have extended this line, it appears I have encountered another DNA match who is further back than any of the cousins mentioned previously.

My 39th Adams First Cousin

The last match came as a surprise to my family. Although I already knew of 38 maternal first cousins, in November 2013 I found another one on my AncestryDNA match page. However, no one in his pedigree chart was familiar to me. I soon discovered that this individual had been adopted as an infant and the pedigree chart he had posted was for his adoptive family. You will read more about this discovery in a later chapter.

FINDING MATCHES AT 23ANDME

The matches you will get at 23andMe will have a different distribution than those you get at FTDNA or AncestryDNA. Most customers of FTDNA and AncestryDNA delved into the world of genetic genealogy because they were genealogists or at least were curious about their ancestry. At 23andMe, many, if not a majority, of those in the database took the test because they wanted to find out something about their health. Often people with this primary motivation are not interested in genealogy or mistakenly think that by pursuing this aspect of family history they will be disclosing something in their health history that they may wish to keep private. 23andMe is very concerned about privacy. For this reason, you must request permission to share genomes and genealogical information with your matches to find genealogically relevant connections or work past brick walls.

23andMe's Large Database

As I was writing this chapter, I got an email from a colleague who is a very experienced genealogist but is a novice at DNA testing. He wrote, "It seems I have gone from famine to feast. 75 new relatives have joined 23andMe in the last 30 days!" He had previously become very discouraged because he had almost no matches of any kind on his yDNA. Now he felt overwhelmed with possible cousins. "This is more than I can cope with. How do I decide what to do?"[12]

My suggestion to him was the following:

1. Don't be too concerned about the number of matches you have. Many of them:
   - Are as unsure of how to proceed as you are
   - Underwent testing for health information and are not really interested in genealogy
   - Know nothing about their family
   - Check their internal messages within 23andMe only very rarely
2. Make sure you respond promptly to internal messages from within 23andMe. Consider trying to move your conversation to your personal email if the match seems promising. However, you may find it easier to keep track of what you know about each match by continuing to keep your communications within 23andMe. As you work with more and more matches, you will need to develop some system for keeping track of each one. Some genetic genealogists like to use a spreadsheet to keep track of their potential cousins as they work with them to identify how they are related.

3. You can investigate some of your matches by viewing their profile pages to see if any helpful information is contained there. The flip side of this is to make sure you have made your profile page accessible to potential cousins and that it has relevant information:

    a. Include your ancestral surnames.

    b. The current pedigree chart at 23andMe is underdeveloped, difficult to navigate, and rarely useful. If you have an online pedigree chart, you can list a link to it if you care to share it with all people who may have access to your profile page.

    c. You can add any other pertinent information that potential cousins might need to be able to evaluate whether it is worth their time to contact you.

4. Assess your first page of "DNA Relatives" under the "Family and Friends" tab. The closest matches should appear at the top of the list.

5. Compose a short paragraph that summarizes your ancestry and provides a link to your personal pedigree chart on your own site or another site like Ancestry.com. Keep it relatively brief and friendly. It is your intent to introduce yourself and to pique the person's curiosity. It is not in your interest to overwhelm someone by force feeding them everything you know about your family. Save this paragraph for refinement and repeated reuse. Copy this introduction so that you are able to paste it into sharing invitations to your best matches.

6. Now you are ready to begin the match making in earnest. Don't be distracted by the hundreds of matches you may have. Start at the top of those listed on the front page of your DNA Relatives list. In most cases, they will be your closest matches and the easiest with whom to find common ancestry. There are exceptions to this, however, as those individuals from endogamous populations (e.g., Ashkenazi Jewish, Puerto Rican, Pacific Islander, or in my case Anabaptist) tend to be related as sixth or seventh cousins on more than one ancestral line, yet can be predicted as a much closer relationship, such as a third cousin, by 23andMe's algorithms.

7. Scroll down the list until you come to the first match who is listed in the right column with a link that is labeled "Send an Introduction." You are about to make a "cold call." If you are fortunate, your matches may have displayed some relevant information in their profile, such as family surnames, locations, and maternal or paternal haplogroup membership. Note this information, but do not get distracted by it unless there are shared surnames or locations that are very specific to your family as well.

8. When you start to send a sharing request, 23andMe will help by providing you with a short paragraph that describes the level of your match. This is a good starting point for your message. Paste your personal paragraph (the one you composed in Step 5) below the 23andMe paragraph. Add an ending such as "What do you know about your ancestry?" Also add your name and at least your general geographic location. Your purpose is to begin a bonding process without overwhelming your match. Recopy and save your message except for what 23andMe composed for you. It is better to let 23andMe describe each match exactly with what the test results can disclose about the closeness of your relationship.

9. Make sure that the option "Share my name and profile and also extend an invitation to share genomes without health information" is selected. As will be discussed later, it is in your interest to be able to share as soon as possible your basic genomic information so you can begin to explore your genealogical connection without the specificity that discloses health information.

10. Send the invitation.

11. Move to the next closest match that is designated with "Send an Invitation." Repeat the process described in Steps 7 through 10. You can now paste your personalized part of your

invitation at the end of 23andMe's description of your matching level and send another invitation with just a few clicks.

12. Continue this process until all your "Send an Invitation" matches on the first page of your DNA Relatives list have been invited to share.

13. Now move to your closest "Public Match" on page 1. With them, your task may be easier because they have already disclosed their names or at least their pseudonyms. However, you may need to tell them how closely you are related because 23andMe will not automatically generate that detail for you.

14. Click on the potential match's name (or pseudonym) on your DNA Relatives page. This will take you to the person's profile page. After you have read any relevant information in the profile, click on "Invite [person's name] to share genomes." Don't get trigger happy and immediately click the send button. Instead, click on "Customize Message." Add the match's name in a salutation to make the message more personal (e.g., "Dear Sally" or "Hi, Sally"). Briefly tell the person about the level of match that 23 and Me believes "Sally" to share with you. Then paste in the paragraph you were using for the "Send Introduction" matches described earlier.

15. As soon as your time and energy level allow, you should repeat these processes at least through the matches found on pages 2, 3, and 4.

These steps will incorporate your closest 100 matches into your genealogical quest and may give you some new cousins with whom you can work to become acquainted. Do not be too concerned that you will be overwhelmed by immediate responses from all 100 people. Some will never respond. A few will decline your request, and you will get a response from some others two or three years from now.

Check back every few weeks to find out how many new "cousins" have tested recently that you should be inviting to join in dialogs about your relationships. The sending of invitations is the easy part, for which a cookie-cutter, by-the-numbers approach may suffice. When your matches respond, the hard genealogical work will begin that will tax every bit of ingenuity that you can muster.

When You Are Contacted by a Match

You may be contacted by matches at 23andMe for either of two reasons:

• They have initiated contact.
• They are responding to one of the invitations you sent.

If they have initiated contact, quickly evaluate whether you were selected at random, selected because of something in your Profile, and/or selected because of an actual matching segment of atDNA. This will help you decide how much time you should invest looking for a genealogical match with this individual.

If the person is responding to your invitation, you already know that the two of you are matches. Now you will have an opportunity to learn more about your match. Until now, unless these people were public matches, you did not know their names. Now you will have an opportunity to learn more. It is now "genealogy game on."

To advance toward identifying your common ancestors, there are a number of steps you can take:

- Read the message the person sent through 23andMe's internal messaging system. Ideally, the person will have at least written a short note.
- Visit or revisit the individual's profile page.
- Select the "My Results" tab from your main 23andMe page, then select "Ancestry Tools," and finally select "Family Inheritance: Advanced." Enter the name of your newly discovered match in the box on the left side of the screen. Enter your name and the names of up to two other previous matches that you think might also match in the three boxes on the right side of the screen and then click "Compare." Carefully note the chromosome location of your matching segments—particularly those segments longer than 10 cM. (If your matching segments should happen to include one on chromosome X, you will find further information on analyzing your match in Chapter 6.) The default display shows a summary comparison of your matches followed by a plot of the matches on an image of your chromosomes (Figure 4.8).

The same information can be displayed in a table format (Figure 4.9). This format is not as visually appealing but it presents a more exact technical description of the matches. You may find it easier to bond with your matches if you copy and paste your equivalent of this table and send it to the person when you thank him or her for sharing.

Figure 4.8

Overlaying the matching segments of two potential cousins on the genome of a known individual

Comparison	Half IBD	# segments
MW vs. DD	60 cM	3
MW vs. JD	63 cM	3

Figure 4.9
Technical details of the matching segments shown in Figure 4.8

Comparison	Chromosome	Start Point	End Point	Genetic Distance	Number of SNPs
MW versus DD	16	54000000	75000000	26.1 cM	4179
MW versus DD	17	32000000	35000000	7.4 cM	829
MW versus DD	17	70000000	81000000	26.6 cM	2292
MW versus JD	16	54000000	76000000	26.9 cM	4315
MW versus JD	17	31000000	35000000	8.5 cM	970
MW versus JD	17	70000000	81000000	27.3 cM	2420

23andMe will let you compare your genome with up to five matches who already have agreed to share genomes. This will allow you to see if all of you share a particular segment or segments in common.

Pay particular attention to the "Genetic Distance"[13] column in Figure 4.9. You should concentrate your efforts on your closest matches unless you are addressing a particular question. Normally you will want to focus on matches with whom your longest block of shared atDNA is 10 cM or greater. Such matches have a 99 percent probability of being inherited from a common ancestor or "inherited by descent" (IBD; Figure 4.10).[14] Shorter matches can also be useful, but investigate them with care. Searching for common ancestors with whom you share a longest match shorter than 8 or 7 cM will quickly lead to diminishing returns and false positives or segments that are "inherited by state" (IBS). However, you will encounter cousins who are obsessing over matches around 3 cM. At this level, the probability that your "shared segment" came from a common ancestor is less than 1 percent. You may wish to avoid getting caught up in those would-be cousins' research frenzy. You should have other matches on which you can make more productive use of your time.

Figure 4.10
Probability of a legitimate match that was "inherited by descent" (IBD)[*]

cM Shared as Longest Block	Probabilities	
	% IBD	% IBS
10 cM	99%	1%
9 cM	80%	20%
8 cM	50%	50%
7 cM	30%	70%
6 cM	20%	80%
5 cM	5%	95%

*"Identical by Descent Segments," *ISOGG Wiki*, http://www.isogg.org/wiki/Autosomal_DNA_statistics#Identical_by_Descent_segments, accessed March 27, 2014. This table is based on the research of John Walden.

The same guidelines apply to "Family Finder" matches at FTDNA. You can find them by selecting the "Show Full View" option on the main Family Finder page. Unfortunately, AncestryDNA does not yet give you your information in this format.

Did That Ancestral Couple Give Us Our Shared DNA?

After I had finished the first draft of this chapter, two different DNA Relatives matches notified me, within a few hours of each other, of shared ancestors in 17th-century Colonial America. Shared atDNA can be inherited from ancestors far in the past. However, as the generations pass, it becomes increasing improbable that long segments will have been inherited intact. In neither case did my son inherit enough of these segments to report as a match.

I share 11 cM of atDNA with the first match. That is just enough to push our probability of having a legitimate match into the more than 99 percent range. However, the shared ancestral couple whom my match had found by meticulously combing through our pedigree charts was married in England in 1601. They never came to the colonies, but at least two of their children did after they, too, had married in England. The parents are my 10th great-grandparents. While it is possible that my match and I inherited this 11 cM segment from another couple who has so far eluded my pedigree chart or that of my match, it may be more probable that we inherited it from our distant English forbearers. We would need to test more descendants to be 100 percent sure about whether this atDNA segment was inherited from an ancestral couple that many generations removed.

As I was examining these matches, Angie Bush referred me to a 2011 blog post by Steve Morse. For those of you who are not familiar with Steve Morse, he is a brilliant software engineer who back in the late 1970s designed the 8086 personal computer chip for Intel. Fortunately, he is also an avid genealogist. His site "One Step Webpages" (www.stevemorse.org) is a treasure trove for the rest of us. From his research on his own 23andMe matches, he has offered these insights:

- Distant relatives often share no genetic material at all.
- It is possible to share a segment with very distant relatives.
- Sometimes, more distant relationships are more likely.
- Most of your relatives may be descended from a small fraction of your ancestors.[15]

You may be interested in reading his post for yourself. Suffice it to say, I found his second and third points to be on target to the matches I was examining.

The second match appears to need more analysis. The ancestral couple was a little more recent, and the matching segment was significantly longer—26.7 cM. Both members of a shared ancestral couple, spotted on my *father's side* of my pedigree chart by my match's eagle-eyed son, were born in the mid-1600s in the Massachusetts Bay Colony. They would be seventh great-grandparents for me and eighth great-grandparents for my match. Based on Morse's statistical calculations, it is possible and may be probable that we both inherited this atDNA from that couple. But wait: this new match has 23andMe results uploaded to GEDmatch. Some of my more immediate cousins have Family Finder results uploaded there. Two cousins also appear to share almost all (25.3 cM and 19.7 cM) of this same segment. They are on my *mother's side* of the family. Oh, how fickle atDNA can be. Our hunt for the donors of our matching DNA goes on.

The moral of these stories is that just because you share atDNA and you find common ancestors, it does not always mean that your shared atDNA came from those ancestors. To be certain, several descendants with a shared segment should be able to trace their ancestry back to a common person.

FTDNA'S FAMILY FINDER

Family Tree DNA's autosomal test is called Family Finder. It offers the following major components:

1. *Your Matches.* Sample screen shots of this feature are included later in this chapter.
2. *Population Finder.* This component will be discussed at the end of this chapter.
3. *Chromosome Browser.* Sample screen shots of this feature are included later in this chapter.
4. *Known Relationships.* You can confirm with your matches that they agree with the relationship you have proposed.
5. *Download Raw Data.* This tool enables you to export your raw data, so that you can then import it into third-party utilities like GEDmatch.com. With FTDNA, this is a two-step process. Autosomal data must be downloaded separately from xDNA data. Remember to upload both files into GEDmatch. Raw data files can be downloaded from 23andMe and AncestryDNA in one step.
6. *Advanced Matching.* This component allows you to get an overview of your matches to see which other tests they have taken at FTDNA.

Family Finder: An Ostfriesland Match?

This next example involves genetic genealogy techniques that are a little more advanced. It may take a little more concentration to follow.

My late father-in-law, William Christie, was known within the family as "Grandpa Willie." He is the individual (born in 1918) in the fourth row, just below the middle in Figure 4.11.

As I explored his matches in FTDNA's Family Finder (Figure 4.12), the closest matches were no surprise.

1. First listed were his two daughters, Michele (Christie) Sanner and Denise (Christie) Dowell, who each shared about 3,384 cM in length or half of Grandpa Willie's atDNA.
2. Then came two grandsons—David "Lee" Smithey and Jason Smithey.
3. Then came a maternal first cousin, Robert McLaren—the son of his mother's brother.
4. The next closest matches, projected to be second to fourth cousins, were with unknown individuals. The first two of these shared a longest block of 44.36 cM with Grandpa Willie. It would be highly unusual if these two women did not share the same 44.36 cM of DNA with each other. This would make them closely related to each other. However, based solely on Grandpa Willie's list of matches, there is no guarantee that both share the same segment.

This list of "Grandpa Willie" Christie's closest matches in Family Finder includes both known close family members and unknown atDNA match "cousins." Names have been removed from the listings of those who have not given explicit consent to be used as examples.

I used the "in common with" tool within Family Finder to explore the extent to which these next closest matches might cluster and match each other (Figure 4.13).[16]

Figure 4.11

Selected ancestors and descendants of William John "Grandpa Willie" Christie II

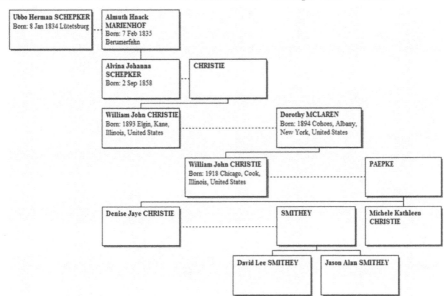

This tool allows you to choose two individuals and search the database for any other individuals who are related to *both* of them. To do this, I selected the closest unknown second to fourth cousin—in this example, Melissa Hawthorne—and created a list of individuals who shared atDNA with both Grandpa Willie and this previously unknown atDNA match.

Three non-immediate family members of Grandpa Willie also had solid matches with both Grandpa Willie and Melissa. (For this exercise, I disregarded a few "in common with" matches whose longest block shared with Grandpa Willie were less than 10 cM. Remember that the probabilities of false positives escalate significantly as the lengths of the longest blocks shared dip through single-digit centiMorgan values.) I then used the Chromosome Browser tool in Family Finder to display the segments that Melissa and each of these other three people shared with Grandpa Willie. I excluded his immediate family members from this comparison because they would have had so many matching segments that it would have been difficult to evaluate the other matches. The two daughters would have had half identical matches across his entire genome and the grandsons would also have had extensive matches. My purpose was to learn about close matches outside the immediate known family. Some of these had clustered results on chromosomes 1 and 2 (Figure 4.14).

As seen at the top of Figure 4.14, all four cousins share overlapping segments on chromosome 2. The top two share what appears to be an exact match with each other as well as with Grandpa Willie. Three of the four share an additional segment on chromosome 1.

While this cluster looks very promising for further analysis, we need to be careful not to get ahead of ourselves. As genetic genealogist Rebekah Canada cautions, "You may all share a DNA segment inherited from the same ancestor. However, you each inherited half of your DNA from each parent. Sharing with any two matches may

Figure 4.12
"Grandpa Willie" Christie's closest matches in Family Finder

be through a different parent."[17] Recall that at each location along our chromosomes, we have atDNA from both our mother and our father. To make sure these segments are from the same side of Grandpa Willie's family, another step is required.

In Family Finder's Matrix view, I loaded all of Grandpa Willie's tested close family members, along with those who were in the newly discovered cluster. The resulting matrix is shown in Figure 4.15.

My wife, Denise Dowell, was the only person in the matrix of Grandpa Willie's matches who also matched all the other individuals. Melissa Hawthorne matched all but Robert McLaren. Note in Figure 4.15 the absence of any matches between Robert McLaren and anyone outside Grandpa Willie's immediate family. While the absence of matches is not conclusive evidence that the matches seen on chromosomes 1 and 2

Figure 4.13

"In common with" matches for Grandpa Willie and Melissa Hawthorne

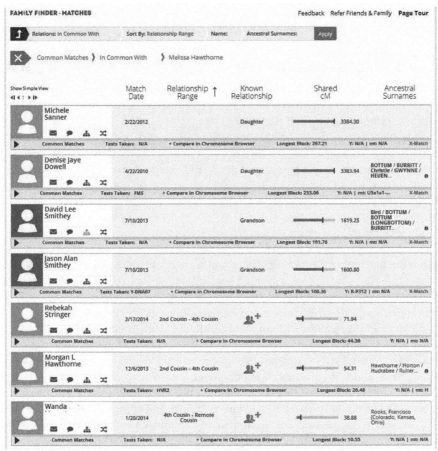

(Figure 4.14) are not related to Grandpa Willie on his maternal side of the family (which he shares with his first cousin Robert), it certainly supports the perception that this is a reasonable hypothesis for further investigation.

After contact was established through the email address provided in Grandpa Willie's match list at FTDNA, Melissa Hawthorne told me that Rebekah Stringer is her daughter, and Morgan Hawthorne is her niece—the daughter of her brother. As I viewed a pedigree chart that Melissa and her niece Morgan had created, it caught my attention that Melissa had a grandmother, Christina Lottmann, who was born in Kleinheide, Ostfriesland, in 1913. The selected Hawthorne descendents of Christina Lottmann who was born in this town in Germany are shown in Figure 4.16.

This finding was significant because Grandpa Willie also had a grandmother, Alvina Schepker (see Figure 4.11), born in this extreme northwest corner of Germany in 1858. We do not know exactly where Alvina as born, but we know that her father, Ubbo Herman Schepker, was born in Lütetsburg in 1834. Lütetsburg is approximately 3½ miles from Kleinheide. Alvina's mother, Almuth Marienhoff, (Grandpa Willie's

Figure 4.14

Clusters of matching blocks of atDNA that four unknown cousins share with Grandpa Willie

great-grandmother) was born in 1835 in Berumerfehn, which is even closer—approximately 1½ miles from Kleinheide.

This is not conclusive proof that Melissa and Grandpa Willie even match at all on these particular lines. However, if they do not, this will be one of the biggest coincidences I have encountered as a genealogist. It certainly merits serious investigation through extant records from Ostfriesland. What we know to date suggests that it is possible that Melissa and Grandpa Willie share a common ancestor as recently as the very early 19th century.

My wife Denise is a match with everyone else in this matrix (Figure 4.15). However, demonstrating the capricious inheritance pattern of atDNA, her sister Michele does not match the apparent outlier Wanda. These sisters share only 44.8 percent of their atDNA. That is less than the 50 percent average but well within the expected distribution of the 37 to 63 percent range of shared DNA for siblings. This example indicates why it is important in atDNA analysis to test as many siblings as possible. While testing more than one sibling generally will not add useful genealogical information when we are using yDNA and mtDNA, that is not the case with atDNA. Down through the second cousin level, their atDNA matches should be the same. However, at the fourth cousin level, each would be expected to match about half of their cousins in that database and most likely not all of the same ones. At the fifth cousin level, each would be expected to match only about 10 percent of the cousins in that database. Unless they are identical twins, each sibling tested uniquely will match yet additional distant cousins. This testing of multiple siblings is not necessary if both or their parents have been

Figure 4.15

Matching status of each individual using the Matrix tool in FTDNA's Family Finder

	Denise Jaye Dowell	Michele Sanner	David Lee Smithey	Jason Alan Smithey	Robert Weig McLaren	Melissa Hawthorne	Rebekah Stringer	Morgan Leigh Hawthorne	Wanda . .
Denise Jaye Dowell		✓	✓	✓	✓	✓	✓	✓	✓
Michele Sanner	✓		✓	✓	✓	✓	✓	✓	
David Lee Smithey	✓	✓		✓	✓	✓	✓		
Jason Alan Smithey	✓	✓	✓		✓	✓	✓	✓	
Robert Weig McLaren	✓	✓	✓	✓					
Melissa Hawthorne	✓	✓	✓	✓			✓	✓	✓
Rebekah Stringer	✓	✓	✓	✓		✓		✓	✓
Morgan Leigh Hawthorne	✓	✓		✓		✓	✓		
Wanda . .	✓					✓	✓		

✓ - This person is identified as a match.

Figure 4.16
Selected Hawthorne descendants of Christina Lottmann

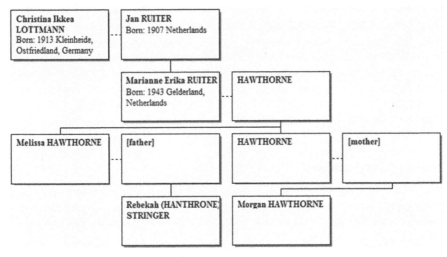

tested. However, although Grandpa Willie was tested before he died in 2011, his wife died prior to the advent of direct to consumer atDNA testing.

CASE STUDY: SOLVING A MYSTERY WITH WOMEN'S DNA

Many people still believe that only male DNA is useful for genealogical purposes. They need to wake up and smell the DNA developments of the last four years. You were introduced briefly to the closely held family secret described in this case study at the end of Chapter 3.

I recently helped a woman unravel an old mystery in her family using her autosomal and mitochondrial DNA test results. I will call her "Allyson" to protect the guilty dead and the living innocents. Here are some of the facts:

- The adopted daughter of Allyson's great-aunt was thought to be a family member.
- This adoptee was born in the 1920s.
- The adoptee's daughter, whom I will call "Bertha," is alive and has taken a mtDNA test and a Family Finder atDNA test at FTDNA.
- Allyson also has taken both an mtDNA test and a Family Finder atDNA test at FTDNA.

Speculation within the family included the following stories:

- This adoptee was the result of an illicit coupling of Allyson's grandmother and her ex-husband long after their divorce.
- This adoptee was the result of a teenage fling of Allyson's mother.

The mtDNA Results

Mitochondrial DNA tests are more definitive in ruling out potential relationships than in proving them. Their results could have ruled out Allyson's mother and grandmother as potential mothers for Bertha's mother. However, Allyson and Bertha are exact matches over all 16,569 locations on their mitochondrial DNA. In mid-2013, they were the only exact matches for each other in the database. This means that the two share a direct maternal line (umbilical cord) ancestress probably within genealogical time. FTDNA says that such a match has a 50 percent probability of a common direct umbilical cord line ancestress within five generations and a 95 percent probability of a common direct umbilical cord ancestress within 22 generations. In this case, the results do not prove how close the common ancestress actually lived, but they do add some credibility to the belief that this was an adoption within the family. Based on this result, some hypotheses were formulated.

Research Hypotheses

1. Bertha's mother was the biological daughter of Allyson's maternal grandmother and grandfather.
2. Bertha's mother was the biological daughter of Allyson's maternal grandmother and some unknown male—at least unknown to us.
3. Bertha's mother was the biological daughter of Allyson's mother and some unknown male.

No hypothesis was defined to include Allyson's mother *and* father being the biological parents, as existing letters clearly document that they did not meet until a few years after the birth of Bertha's mother. Conversely, it can be documented that Allyson's maternal grandmother and grandfather remained in contact long after their divorce.

Dick Eastman has famously blogged that there is no such thing as a half cousin.[18] He quoted the venerable *Black's Law Dictionary* to support his assertion. Well, Dick, we would never be able to solve some of our genealogical mysteries unless we were able to split some cousins and other relatives in this manner.

If we want support for Hypothesis 1, we would need to find evidence that Allyson and Bertha could be first cousins. To support Hypothesis 2, we would need to find evidence that Allyson and Bertha could be half first cousins. If Hypothesis 3 is correct, Allyson would have a half aunt relationship with Bertha. In these three cases, Allyson and Bertha would be expected to share either about 12.5 percent of their atDNA (first cousins), about 6.25 percent (half first cousins), or 12.5 percent (half aunt) (Figure 4.18). These percentages can be derived from the chart found on the ISOGG wiki.[19]

Oops! Using the autosomal data at hand, we would not be able to distinguish between Hypotheses 1 and 3. The expected amount of shared DNA would be the same in both cases. Fortunately, there are three living first cousins of Allyson who might be willing to be tested. Two are female and one is male. In this case, their sex would not matter: all we would need is their autosomal matches with Allyson and Bertha. These cousins would not add to our ability to distinguish between Hypotheses 1 and 2 because their amounts of DNA shared with Bertha should mirror those of Allyson in both cases. Both Allyson and each of the three cousins should be shown to be first cousins with Bertha in Hypothesis 1 and half first cousins in Hypothesis 2. Thus our diagnostic ability is not increased by collecting DNA from the three additional cousins.

Figure 4.17
Cousin tree with genetic relationships

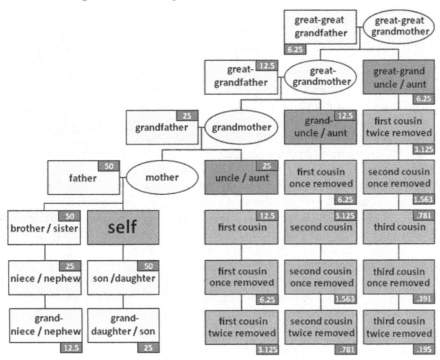

But wait! We do not need them to help distinguish between Hypotheses 1 and 2. We might need their test results to differentiate between Hypotheses 1 and 3, however. Great! While Allyson would be a half aunt (sharing about 12.5 percent of her atDNA)

Figure 4.18
Expected percentage of shared atDNA for each hypothesized relationship

Relation with Adoptee's Daughter "Bertha"	Maternal Grandmother Is Mother of Adoptee		Allyson's Mother Is Mother
	Maternal Grandfather Is Father (Hypothesis 1)	Maternal Grandfather Is Not Father (Hypothesis 2)	Allyson's Father Is Not Father of Adoptee (Hypothesis 3)
"Allyson"	First cousin—12.5%	Half first cousin—6.25%	Half aunt—12.5%
Male cousin	First cousin—12.5%	Half first cousin—6.25%	Half first cousin once removed—3.125%
Female cousin 1	First cousin—12.5%	Half first cousin—6.25%	Half first cousin once removed—3.125%
Female cousin 2	First cousin—12.5%	Half first cousin—6.25%	Half first cousin once removed—3.125%

in Hypothesis 3, the other three cousins would only be half first cousins (sharing about 3.125 percent of their atDNA)—a significant difference. It looks like it may be possible to collect enough information to solve this mystery.

The Solution

Actually, the solution turned out to be quite a bit simpler than all that. As we saw much earlier in this saga, FTDNA predicted that Allyson and Bertha were first to second cousins. For our purposes, that is quite an extensive range. If they share the matching segments that would be expected of first cousins, they would also share the amount of atDNA Bertha would expect to share with a half aunt. But that is not the amount they share—they share only half that amount. In this case, it would have been easier to determine this fact if Allyson and Bertha had been tested at 23andMe rather than at FTDNA, because the former reports matches in terms of percentage of match, while the latter reports these data in terms of centiMorgans (cM) and SNPs matched.

Figure 4.19 shows the *average* percentage of autosomal DNA shared with relatives, assuming that every child gets 50 percent from his or her mother and 50 percent from his or her father and the average number of shared centiMorgans.[20] However, the actual amount will vary somewhat.

When interpreting Figure 4.19, be careful to remember that the percentages and centiMorgans are *averages*. In actual practice, you will find that the nice clean categories shown in this figure get messy and sometimes overlap. One version of this variation may be seen on the ISOOG wiki.[21]

Figure 4.19
Average amount of atDNA shared with various relatives

Percentage	centiMorgans	Relationship
50%	3400.00 cM	Mother, father, siblings
25%	1700.00 cM	Grandfathers, grandmothers, aunts, uncles, half-siblings, double first cousins
12.5%	850.00 cM	Great-grandparents, first cousins, great-uncles, great-aunts, half-aunts/uncles, half-nephews/nieces
6.25%	**425.00 cM**	**First cousins once removed, half first cousins—best answer for our data**
3.125%	212.50 cM	Second cousins, first cousins twice removed
1.563%	106.25 cM	Second cousins once removed
0.781%	53.13 cM	Third cousins, second cousins twice removed
0.391%	26.56 cM	Third cousins once removed
0.195%	13.28 cM	Fourth cousins
0.0977%	6.64 cM	Fourth cousins once removed
0.0488%	3.32 cM	Fifth cousins
0.0244	1.66 cM	Fifth cousins once removed
0.0122%	0.83 cM	Sixth cousins

In our case study, Figure 4.19 indicates that the 428.42 cM shared between Allyson and Bertha is well within the expected range that would be estimated for a "first cousin once removed" relationship. But first cousin once removed is not one of the expected outcomes given in the hypotheses. Or is it? Isn't it the equivalent of a half first cousin? I think we have a winner! Hypothesis 2 best fits the autosomal results. The most likely biological parents of Bertha's mother are Allyson's maternal grandmother and an unknown male. Allyson's maternal grandfather seems to be ruled out as the biological father of Bertha's mother. Although testing additional cousins might lead to greater certainty, that does not appear to be necessary to identify the most likely candidate to be Bertha's grandmother.

Most of our discussion in this chapter has been about situations in which we have been fairly passive in our analysis of matches that have come our way. We went fishing for matches in various databases and then tried to find meaning in those we found. This is what Maurice Gleeson called efforts "to connect with distant cousins." Only in the situation of Allyson and Bertha have we taken control of the situation and set up hypotheses to deliberately focus on solving a long-standing family mystery. In so doing, we exploited two of Gleeson's purposes for atDNA testing—"to confirm known or suspected relationships" and thereby "break through a brick wall." As you become more familiar with genetic genealogy, you will encounter more situations where you can be more proactive and carefully construct tests that might potentially help you solve your own genealogical riddles.

This leaves us with Gleeson's fourth purpose for atDNA testing, "to determine your ethnic makeup." That topic will be explored next.

WHERE DID MY ANCESTORS COME FROM?

So from where did your ancestors come? As Maurice Gleeson reminded us earlier in this chapter, many people take atDNA tests to determine their ethnic makeup.[22] For some individuals, this may be their prime motivation. If this is the case, it helps build our genetic genealogy databases so we all benefit, unless those persons seeking their homelands become disillusioned by their admixture results. Admixture predictions are an area where the application of genetics is on less than solid ground. It is not that the genetics is at fault. Rather, it is the inexact knowledge of where particular populations have lived within the various time/location continua. Fortunately, we are assisted by population geneticists, anthropologists, and archaeologists, among others, as we climb this steep learning curve. These related fields are progressing quickly, though much is still not known.

Location, Location, Location

When I present genealogy seminars, I am often asked where my ancestors came from. My response is to ask which time period the questioners are thinking about. If they are thinking of the last three centuries, my ancestors are almost entirely native to America.[23] Certainly, all of them who were born in the last two centuries discovered planet Earth here in North America. If my inquirers are thinking about 500 to 1,000 years ago, my ancestors were European.

Over the last 500 years most of our DNA has passed down through fifteen to twenty generations. It all depends whether we use 25 years per generation or 33 years as the average generation cycle. In fifteen generations we could have had more than thirty-two thousand discrete ancestors. In twenty generations our number of potential ancestors balloons to more than a million. Because of inbreeding our actual number of discrete individual ancestors is somewhat less. However, the number of ancestors who could have contributed atDNA to our genmomes over the last five centuries is quite large. Of course not all of them have contributed atDNA that has survived the random inheritance patterns and made it down to us. So it is not realistic to expect an atDNA test to be able to tell us where our ancestors were 500 years ago. Which of our thousands of potential ancestors are we expecting to find?

Based on yDNA and mtDNA I know that about 10,000 years ago, most of my ancestors were West Asians, although some may have been in the lands that border the Mediterranean. If they are thinking about 50,000 years ago, my ancestors were mostly African. In many respects, this is not a satisfactory answer. However, it illustrates some of the difficulty of expecting our DNA to tell us from whence we came. The various segments of my DNA have come together after experiencing wide-ranging journeys across four continents. Yours may have followed a similar circuitous path.

Rule 8 in my *Crash Course in Genealogy* is "Look for records where they would have been recorded when they were created. Location, location, location! Location in place, location in time, location in records repositories of the governing body at the time of originating time and place."[24] Our DNA records repositories are very complex, particularly when it comes to telling us where they have been along a particular sector of their time continuum. This kind of genealogy record repository was much more highly mobile than other records we are accustomed to consulting in our research. Our human cells, even though they wrote a travelogue with their SNPs, sometimes skipped eons between diary entries. In addition, these entries were not made in a tidy linear timeline—at least not one that we have completely deciphered.

Neanderthal Ancestry?

Buried among these diary entries was some information that caused me to change my image of my ancient ancestors. Approximately 20,000 to 40,000 years ago, many of my ancestors had become Caucasians—living in the vicinity of the Caucasus mountains in Central and West Asia. At that time, at least one of my ancestral lines was still in Africa working its way to the shores of the Mediterranean. Approximately 50,000 years ago, all my ancestors, with the potential exception of a few stray Neanderthals and their Denisovan cousins, were living in Africa.

Who and what were these Neanderthal and Denisovan ancestors? The National Genographic Project (Geno 2.0) tries to help us answer this question:

> When our ancestors first migrated out of Africa around 60,000 years ago, they were not alone. At that time, at least two other species of hominid cousins walked the Eurasian landmass—Neanderthals and Denisovans. As our modern human ancestors migrated through Eurasia, they encountered the Neanderthals and interbred. Because of this, a small amount of Neanderthal DNA was introduced into the modern human gene pool.
>
> Everyone living outside of Africa today has a small amount of Neanderthal in them, carried as a living relic of these ancient encounters. A team of scientists comparing the full genomes of the two species concluded that most Europeans and Asians have between 1 to 4 percent Neanderthal DNA. Indigenous sub-Saharan Africans have no Neanderthal DNA because their ancestors did not migrate through Eurasia.[25]

The 2.5 percent of Neanderthal atDNA that 23andMe reports me to contain places me among the 22nd percentile among European users, according to 23andMe. My wife and her sister clearly outrank me in this regard. They are reported to be 3.0 percent Neanderthal—placing them in the 91st percentile of those with predominately European recent ancestry.[26] The Geno 2.0 project measures the atDNA of both the Neanderthal and the Denisovan species. According to the latter test, I am only 1.8 percent Neanderthal, but I am also 2.1 percent Denisovan—a category that 23andMe did not single out for reporting. Given that these two species have intersecting origins, some of the Neanderthal counted by 23andMe may have overlapped the Denisovan counted by Geno 2.0.[27]

> Svante Pääbo—who is the world expert in neanderthal [*sic*] and denisovan [*sic*] genomics; his lab is the one who sequenced them—made an announcement that with the new denisovan genome they have they're now reevaluating what a denisovan is because it seems to be about 20% neanderthal. And the fact is that we don't really know what these things are, and the extent to which there is overlap and interbreeding between neanderthals and denisovans.[28]

So I am an American, European, Asian, or African, depending on when you look. The interesting point is not simply where we are now or where we can document the locations of our ancestors in the last few generations; it is our entire human journey. And an impressive journey it has been. My ancestors and yours as well have persevered and survived against incredible odds to get us to where we are today.

Am I Caucasian?

Often times we get surprises when we do our admixture analysis. Except for the Neanderthal and Denisovan estimates, my biggest surprises were some small traces of American Indian, North African, and Middle East/Caucasus ancestry. Although I got considerable variety in my reports from the different companies that tested me, I am considered to be predominately European. The smaller amounts will be interesting to try to explore if they hold up as the underlying populations on which they are based are studied more completely. Being a DNA test-taking junkie, I have ancestry predictions from four companies (Figure 4.20).

Figure 4.20

Ancestral origins projections for author from four different testing companies

Ancestral Origins	European	Middle Eastern/ North African	South Asian	Sub-Saharan African	American Indian
FTDNA[*]	94.34%	5.66%			
23andMe	99.2%	0.4%	0.2%	0.1%	
Ancestry DNA[*]	93%	6%			1%
Genographic 2.0 (predicting many thousand years earlier than the other companies)	41% (Northern European)	38% (Mediterranean)	20% (Southwest Asian)		

*These figures from FTDNA were taken from the old Population Finder. which was still in use in early 2014.
*AncestryDNA's admixture predictions as of April 3, 2014.

The big estimates at the continental level are pretty consistent, but most of us want more. There is danger in wanting more, however, as the science is not yet to a point where these types of specifics can be given. A general rule of thumb is that the more conservative and broad an ethnic estimate is, the more accurate, but less fun, it is. A lot of the action now is trying to estimate our countries of origin back to about 1600 AD or the beginning of a document trail some of us can follow for some of our lines. Many Americans have not been able to find the footprints of some of our ancestors before they got on the boats to cross the Atlantic. We would be happy to get any solid clues as to where to start looking in Europe. Of course, what we would really like is postal zone-specific location information. We may never get that kind of predictive specificity from our atDNA. If we do, it will be far in the future.

Many of us search eagerly for some diversity in our family trees. I still remember my own excitement when I discovered that I had a Swedish ancestor. He was another white man but at least he was not from England or Wales. If you are seeing diversity in your tree, some smaller bits of information in your admixture analysis will be interesting. You can download your raw atDNA data and upload it to third-party site like GEDmatch, then analyze your data to your heart's content. I have raw data there from AncestryDNA, FTDNA, and 23andMe. Each set of source data yields slightly different percentages on the same third-party test. In addition, each test gives slightly different results as it analyzes your data. However, the overall results should be very similar. If you are looking for some small part of your genome that may give evidence to support cherished family myths such as the illusive Cherokee princess or an African American ancestor, you can find many tools to analyze your entire genome or allow you to focus on individual chromosomes. The different results are generally caused by each company's proprietary choice of a slightly different pattern of your genome to test and compare against its proprietary reference populations.

Ancestry Estimates

All of the major labs try to use their clients' atDNA samples to predict the origins of their ancestors. There is a wide variety in the era that each targets, however. Some aim at 500 years ago, while Geno 2.0 aims at thousands of years ago. Obviously, these different time dimensions will lead to different predictions. Our ancestors traveled around the globe over thousands of years.

Blogger Judy Russell, writing in late 2013 about the then state of admixture estimates, had this to say:

> *The Legal Genealogist* has one thing to say about these deep ancestral percentages: *Forget them.*
>
> They're *cocktail party conversation pieces*—and little more. The science just isn't there yet to back them up.
>
> And because of the fact that the only way to get these percentages is by comparing folks like you and me—alive today—to the test results of other people who are alive today (and not to the actual DNA of our ancient ancestors!!), the science may never really be there.[29]

I am not quite as pessimistic as Judy, but I would not want the credibility of genetic genealogy as a whole to be tarnished by the embryonic status of this facet of the field—even though I must admit that looking at these test results is a lot of fun.

Davidski, a blogger and specialist in European population genetics and modern physical anthropology, reminds us of the different precision levels for different population groups around the world: "It's useful to keep in mind that these tests will differ in their interpretation of the data, and perhaps accuracy, depending on the ancestry of the user."[30] When reference samples come from the regions where your ancestors came from, the test results will be more accurate than when they do not. It is useful to remember that most of the people who have taken atDNA tests are from the English-speaking parts of the world, and a majority of them are from the United States. This skewing is not intentional. It results from marketing patterns for DNA test kits and from the regulations on and costs of shipping DNA samples internationally.

Do not get blinded by the apparent scientific precision of admixture reports. Extending these results to the hundredths of a percent gives them the appearance of scientific truth. However, some of you will remember the caveat popularized by Mark Twain when he described the three kinds of lies: lies, damned lies, and statistics. This is a phrase describing the persuasive power of numbers, particularly the use of statistics to bolster weak arguments. It is also sometimes colloquially used to doubt statistics used to prove an opponent's point.

In addition, the seductive power of colorful graphics can lure in the unsuspecting. I am hooked, I admit—but I also take along a shaker of salt when I examine these elements. Maybe several grains would be sufficient.

But do not abort your documentary search for distant ancestral roots just yet. Ideally, all of your information will soon point in the same direction. As admixture analysis matures, it may realize more of its exciting potential. Just remember that we must synchronize a location in time with a location on the planet if we are to have success in using admixture tools to hone in on where our ancestors lived in the past.

The good news is that our atDNA does not have to be retested as new discoveries are made in related fields. The labs can and do recalibrate their admixture predictions as new knowledge about underlying populations is discovered.

NOTES

1. Maurice Gleeson, "Autosomal DNA: A Step-by-Step Approach to Analyzing Your atDNA Matches," presented at Who Do You Think You Are Live, London, UK, February 20, 2014, http://www.youtube.com/watch?v=Jtpe6u2J5ps, accessed February 26, 2014.

2. Rhondalee Lucas, *Lucas & Cummings Heritage by Aussie Rhonda*, http://www.dixie9.com/p/dna.html accessed February 26, 2014.

3. Graham Coop, "How Many Genomic Blocks Do You Share with a Cousin?," *The Coop Lab: Population and Evolutionary Genetics*, University of California–Davis, posted December 2, 2013, http://gcbias.org/2013/12/02/how-many-genomic-blocks-do-you-share-with-a-cousin/, accessed April 2, 2014.

4. James Larry Vick, "Family Finder Price Increase," post to the ISOGG Project Administrator's listserv, April 15, 2010.

5. FTDNA, "Family Finder: Questions about Family Finder," https://www.familytreedna.com/FAQ/answers/17.aspx#628, accessed April 20, 2010.

6. 23andMe, "FAQs: I Know That a Particular Person Is My Relative. What's the Probability That We Share a Sufficient Amount of DNA to Be Detected by Relative Finder?", https://www.23andme.com/you/faqwin/rfprobability/, accessed May 10, 2010.

7. Blaine Bettinger, "The AncestryDNA Witch Hunt," *The Genetic Genealogist,* posted February 16, 2014, accessed February 26, 2014.

8. Debbie Kennett, "The AncestryDNA Witch Hunt," *The Genetic Genealogist,* posted February 16, 2014, accessed February 26, 2014.

9. Maurice Gleeson, "The AncestryDNA Witch Hunt," *The Genetic Genealogist,* posted February 16, 2014, accessed February 26, 2014.

10. Ancestry Help, "Ancestry Hints," http://help.ancestry.com/app/answers/list/kw/Leaf/search/1, accessed February 26, 2014.

11. At this writing, FTDNA is accepting transfers from AncestryDNA. Transfers from 23andMe for kits tested before November 2013 are also accepted. A chip change at 23andMe in late 2013 has rendered its raw data incompatible with FTDNA's processing algorithms—at least for the short term.

12. Del Chausse, email to the author, March 3, 2014.

13. Note that 23andMe and FTDNA use the term "genetic distance" differently. See the Glossary at the end of this book for more details.

14. "Identical by Descent Segments," *ISOGG Wiki,* http://www.isogg.org/wiki/Autosomal_DNA_statistics#Identical_by_Descent_segments, accessed March 27, 2014. This table is based on the research of John Walden.

15. Steve Morse, "Genetic Genealogy and the Single Segment," *On Genetics* (February 19, 2011), http://ongenetics.blogspot.com/2011/02/genetic-genealogy-and-single-segment.html?m=1, accessed April 5, 2014.

16. At this writing, it is necessary to select "Show Full View" just above the first listed match. The default position is "Show Simple View."

17. Rebekah Canada, "Is It Meaningful When Two or More of My Matches Have a DNA Segment or Segments in the Same Location?", *The Family Tree DNA Learning Center BETA,* https://www.familytreedna.com/learn/autosomal-ancestry/universal-dna-matching/meaningful-two-matches-dna-segment-segments-location/, accessed April 2, 2014.

18. Dick Eastman, "There Is No Such Thing as a Half-Cousin!," *Eastman's Online Genealogy Newsletter* (September 27, 2010), http://blog.eogn.com/eastmans_online_genealogy/2010/09/there-is-no-such-thing-as-a-half-cousin.html, accessed March 5, 2014.

19. ISOGG, "Cousin Tree (with Genetic Kinship)," http://www.isogg.org/w/images/4/4c/Cousin_tree_(with_genetic_kinship).png, accessed March 5, 2014.

20. ISOGG, "Simple Mathematical Average of Sharing," http://www.isogg.org/wiki/Autosomal_DNA_statistics#Simple_mathematical_average_of_sharing, accessed March 27, 2014.

21. ISOGG, "Ranges of Sharing Percentage," http://www.isogg.org/wiki/Autosomal_DNA_statistics#Ranges_of sharing_percentage, accessed March 27, 2014.

22. Gleeson, "Autosomal DNA."

23. At least 24 of my 32 great-great-great grandparents were already in the United States 200 years ago. The other 8 may have been, but I have so far been unable to document who and where they are.

24. David R. Dowell, "Dr. Dave's Top 10 Rules for Successful Genealogical Research," in *Crash Course in Genealogy* (Santa Barbara, CA: Libraries Unlimited, year?), 14, 30–31.

25. The Genographic Project, "Why Am I Neanderthal?", https://genographic.national geographic.com/neanderthal/, accessed December 5, 2013.

26. For more information, see Eric Y. Durand, "Neanderthal Ancestry Estimator," *23andMe White Paper* 23-05 (January 8, 2012), https://23andme.https.internapcdn.net/res/pdf/hXitekfSJe1lcIy7-Q72XA_23-05_Neanderthal_Ancestry.pdf, accessed December 5, 2013.

27. Spencer Wells, responding to an audience question at the Genealogy Jamboree, June 2013.

28. Grant Brunner, "Discussing the Personal Genomics Revolution with Nat Geo's Dr. Spencer Wells," *ExtremeTech* (November 7, 2013), http://www.extremetech.com/extreme/ 168586-discussing-the-personal-genomics-revolution-with-nat-geos-dr-spencer-wells, accessed November 7, 2013.

29. Judy Russell, "Those Pesky Percentages," *The Legal Genealogist* (October 27, 2013), http://www.legalgenealogist.com/blog/2013/10/27/those-pesky-percentages/, accessed October 27, 2013.

30. Davidski, "Updated Eurogenes K13 at GEDmatch," *Eurogenes Genetic Ancestry Project* (November 21, 2013), http://bga101.blogspot.com.au/2013/11/updated-eurogenes-k13-at -gedmatch.html, accessed March 28, 2014.

5

What Is the X Factor? A Different Inheritance Pattern

Everyone has X-chromosome DNA. You inherited some from your mother. If you are a woman, you also inherited some from your father. If you are researching your maternal lines, it is great to have a match on your X chromosome—particularly with a male cousin. This will immediately narrow down your potential matching zone to the maternal side of his family. If you are also a male, this will narrow down your potential matching zone even more.

In 2013, I found my first match in which the X chromosome played a significant role. It was the easiest DNA match puzzle I have unraveled so far.[1] One of my daughters-in-law was reported by 23andMe to share a considerable amount of DNA with an unknown male. Her DNA Relatives page predicted him to be a second cousin and reported her to share 3.57 percent of her autosomal DNA with him over 12 matching segments. Unknown second cousin matches are exciting to genetic genealogists. If the prediction is correct, the two individuals shared a set of great-grandparents. That amount of shared DNA could also be consistent with other relationships. The most likely alternative would be a first cousin twice removed.

After the man agreed to share his genome at a basic level (without health information), their match could be explored further. When I compared the two of them in Ancestry Tools—Family Inheritance: Advanced, I could see their matches plotted as an overlay on her chromosomes (Figure 5.1). This gave more good news: my daughter-in-law not only shared 12 segments with the projected second cousin, but their largest shared segment is on their X chromosome. In Figure 5.2, that later segment is shown to be 62.4 cM.

Any of the 12 segments could have been interesting to gung ho genetic genealogists. We love to see these kinds of matching segments between two cousins. Even the two relatively short segments on chromosome 16 are interesting. They are so close together that we could speculate they might have been part of a much larger continuous segment just one intergenerational transfer back in time. However, with the amount of shared DNA in this case, such speculation would have been an unfortunate distraction from much more important findings.

Figure 5.1

The projected second cousins share 12 segments of DNA, including a large match on chromosome X

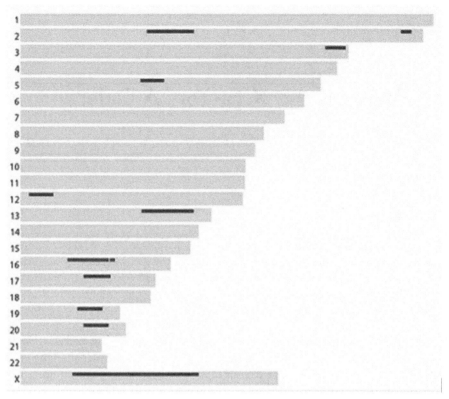

The most noteworthy of these points is that the longest match of all is on the X chromosome. Why is the location of this match such a big deal, you might ask? Recall that the unknown potential second cousin was a male, which means he inherited his entire X chromosome from his mother. We can ignore his father's pedigree chart as we search for the most recent common ancestor. This allows us to cut in half the number of branches of his tree we need to consider, although it did not reduce the amount of the tree of my daughter-in-law that needed to be examined. She had inherited half of her X-chromosome DNA from her mother, but she also had inherited half from her father. As an aside, you can hope someday you will find such a lengthy match between two men in your tree.

I sent a list of my daughter-in-law's four sets of great-grandparents to the wife of her match, who was the family genealogist. The wife responded, "It's Homer and Hilda." Mystery solved! The mother of my daughter-in-law is the first cousin of the mother of the DNA match. The two mothers had lost track of each other after being close in childhood. All DNA matches should be this easy as we seek to identify those who will help us fill in gaps in our family trees!

The shaded boxes in Figure 5.3 represent ancestors who could have contributed xDNA to the two second cousins. Note that there are no father-to-son transmissions.

Figure 5.2

Technical details of the 12 matching segments shared by projected second cousins

Chromosome	Start Point	End Point	Genetic Distance	Number of SNPs
2	77000000	105000000	16.7 cM	3,420
2	231000000	237000000	12.0 cM	1,462
3	185000000	197000000	25.4 cM	2,522
5	73000000	87000000	15.1 cM	2,635
12	5000000	19000000	23.0 cM	3,197
13	74000000	105000000	31.8 cM	6,786
16	28000000	53000000	10.0 cM	1,498
16	54000000	57000000	7.4 cM	1,023
17	38000000	54000000	14.4 cM	2,748
19	34000000	49000000	20.8 cM	2,522
20	38000000	53000000	26.3 cM	3,763
X	31000000	108000000	62.5 cM	5,732

Can you trace the route of the shared segment down two parallel paths from Homer and Hilda to the second cousins?

While my daughter-in-law inherited half of her xDNA from her father, her father inherited his xDNA only from the maternal side of his ancestry (Figure 5.4). A woman inherits no xDNA from her paternal grandfather's quarter of her pedigree chart. She also inherits no xDNA from her maternal grandfather's paternal side, therefore losing any contribution from another one-eighth of her ancestors.

We do not have enough information to determine whether the segment shared by Kindra and Eric came through Homer or Hilda. However, it came from one of them. It was then passed down to their daughters Dolores and Margarie, perhaps originally

Figure 5.3

Example of ancestors who could have contributed xDNA to second cousins

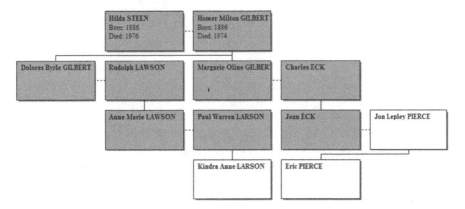

X Chromosome Match Between Second Cousins

Figure 5.4

Approximate percentage of xDNA inherited by a female through each ancestor

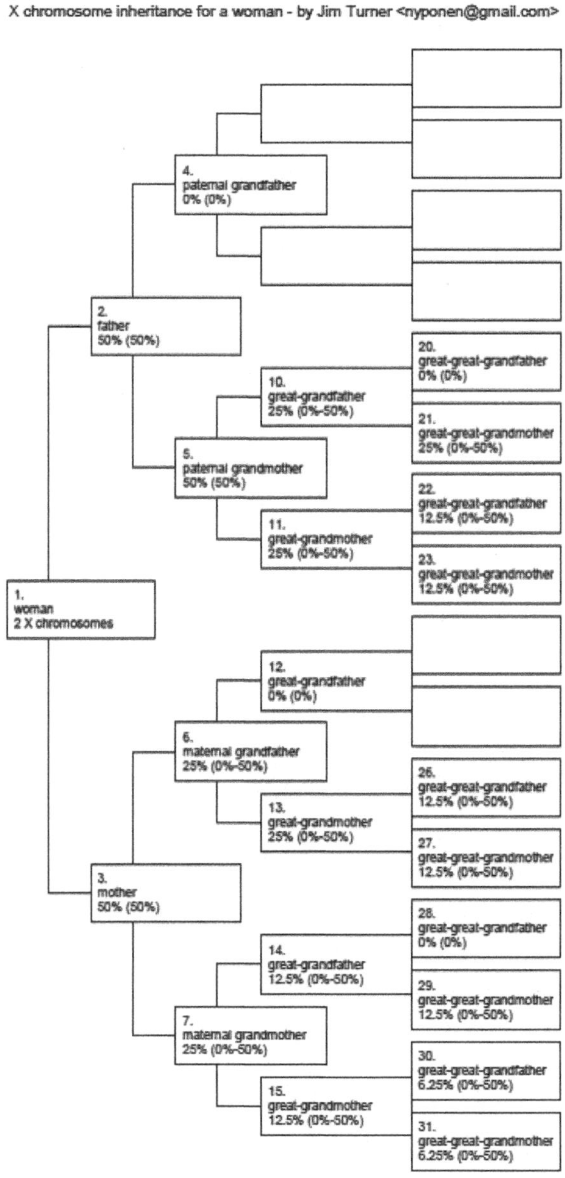

X chromosome inheritance for a woman - by Jim Turner <nyponen@gmail.com>

as a longer segment. Dolores passed her version of the segment down through her daughter Anne and then to Kindra. Margarie passed her version of the segment down through her daughter Jean to Eric.

Figure 5.4 provides the approximate percentage of xDNA that females inherit through each ancestor. Note that as you go back a generation, the average contribution of each ancestor is generally halved. However, *a father-to-son relationship blocks the inheritance of xDNA*. When that happens, the son inherits all his xDNA from his mother.

For men, the loss of the ancestral contribution to their xDNA is even more dramatic (Figure 5.5). They get no xDNA from their father's side of the family. They also inherit none from their maternal grandfather's paternal side, therefore losing any contribution from another one-eighth of their ancestors.

The simple rule at play here is that *a father cannot pass down xDNA to his sons.* Anywhere in a pedigree chart where a male is located, he can inherit xDNA only from his mother. Any xDNA further back from a father-to-son combination is not passed down to the son.

Figure 5.5

Approximate percentage of xDNA Inherited by a male through each ancestor

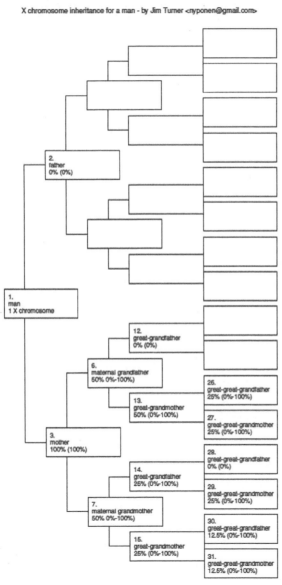

X chromosome inheritance for a man - by Jim Turner <nyponen@gmail.com>

The figures in James C. Turner's charts show the average percentage of xDNA that would be expected to be inherited by males or females. Due to the random inheritance patterns of recombined DNA, the actual percentage would vary within the ranges shown in parentheses in Figures 5.4 and 5.5.[2]

If you are having a hard time grasping these asymmetrical patterns of inheritance, do not despair. Just remember to review them in charts similar to those in Figures 5.4 and 5.5 when you need to trace how a matching segment of xDNA could have been inherited by each individual under analysis. Remember, these charts tell you which ancestors *could* have contributed to your xDNA. You may not have received xDNA from all of these ancestors.

All of us need such reminders from time to time—and that includes your author. At the risk of being redundant, I will assume that repetition is the key to learning. Recently, when I was trying to plug a new first cousin into my family tree, I ignored the four simple rules of xDNA inheritance:

1. Girls inherit xDNA from their mothers.
2. Girls inherit xDNA from their fathers.
3. Boys inherit xDNA from their mothers.
4. Boys *do not* inherit xDNA from their fathers.

Of these four rules, the last is the most important. Anywhere on your pedigree chart where there is a father-to-son relationship, *no* xDNA flows. Instead, yDNA flows. As frustrating as this last rule can be sometimes, it can also be part of the solutions for some of our genealogical mysteries. But like most clues, *they help us only when we are paying attention.*

ANOTHER ADAMS FIRST COUSIN

In December 2013, I blogged about a newly discovered 39th Adams first cousin.[3] I have never had a shortage of cousins—particularly on my maternal side of the family. My mother was one of 14 Adams siblings (10 girls and 4 boys) who survived infancy. A fifth boy died within a few days of birth. My mother was the last surviving sibling when she died in 1998. It was interesting that of the 39 offspring these 14 siblings produced, 18 of them were from the two oldest children. I think the timing of the Great Depression may have had as much to do with this difference in productivity as birth order, but that is just speculation on my part.

When I logged in to AncestryDNA in November 2013, I was surprised to find the following result on my DNA page:

1st Cousin Possible range: 1st-2nd cousins Confidence: 99 percent

The pedigree chart posted by this individual did not help me place the cousin among the 25,000 or so individuals I currently have in my family tree database. The DNA information did not tell me whether this match was on my maternal or paternal side. However, the geographic locations in the pedigree chart of the "new" cousin strongly suggested this match was on my mother's side.

I sent an inquiry through the Ancestry message system and got this response from Jim:

Figure 5.6

Author's connection to his four maternal uncles and his first cousin once removed, Merrill Adams

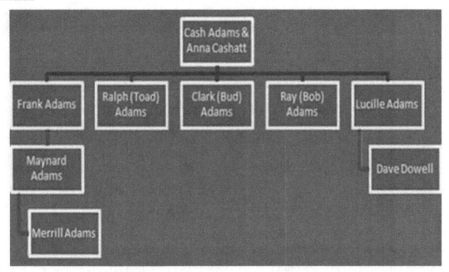

I believe ancestry.com is correct, but the exact connection may be difficult to pin down. I was born 27 September 1932, and adopted at birth by the parents in my ancestry.com family tree. The doctor who delivered me was Dr. G. W. Carpenter, who I believe had an office in Utica, Missouri, at the time. Most of the people who had first-hand knowledge of my birth are now gone. I was reluctant to make inquiries that might embarrass anyone. A neighbor who grew up near Breckenridge vaguely recalled that my father was an Adams employed in a grocery store in Breckenridge, and my birth mother was a young employee. As time permits, I'll try to find out more.[4]

Only one other family member had been DNA tested to my knowledge. That was a first cousin once removed, Merrill Adams, who agreed to be tested so that I could establish the paternal haplogroup of my maternal (Adams) grandfather (Figure 5.6). My DNA sample establishes my paternal (Dowell) grandfather's Y chromosome. Merrill's DNA had also been processed through the FTDNA Family Finder test.

It occurred to me that Merrill's test results could confirm that Jim was a cousin on my maternal side and that Jim would be Merrill's uncle if Merrill's grandfather, my Uncle Frank, was also Jim's father. If instead Jim's father was one of the other Adams brothers, Merrill would be Jim's first cousin once removed. That is the same relationship I have with Merrill.

Since Jim had tested at AncestryDNA and Merrill at FTDNA, the data for both had to be placed in the same database to analyze the nature of their relationship. Therefore, Jim downloaded the raw data from his test at AncestryDNA and uploaded it to GEDmatch.com, a free site that, among other features, allows autosomal results from a variety of labs to be compared with each other. I downloaded Merrill's raw data from FTDNA and uploaded it to GEDmatch.com.

When Jim's results were compared with Merrill's, the two men shared about the same amount of DNA as I shared with Merrill as a first cousin once removed. An uncle would have shared much more with a nephew. Therefore, my Uncle Frank is also Jim's Uncle Frank and is eliminated as a candidate to be Jim's father.

Our search then focused on the three Adams brothers other than Frank. A child of one of the three (who is also one of my previously known Adams first cousins) agreed to take a DNA test. When the test came back, she shared the right amount of DNA with Jim (and me) to be a first cousin. Of the remaining two possible fathers, one never married and has no known descendants. We were able to trace down a grandson of one of the other possibilities to see if he would be willing to take a DNA test. However, we never made direct contact.

At the time, I was operating with blinders on. Those blinders were a couple of vague recollections some of the local old-timers had that an Adams boy may have had an illegitimate child that could have been my new cousin Jim.

In my book *Crash Course in Genealogy*, Rule 4 for genealogists is "Believe everything and believe nothing you hear or see in print."[5] In my first efforts to unravel the mystery of Jim's birth, I was following the "believe everything" part of this rule but ignoring the "believe nothing" part. DNA does not lie. However, sometimes our vision to read the information it contains is considerably less than 20/20.

While I was waiting for additional DNA tests to help determine which of my Adams uncles was Jim's father, I decided to mount a second investigation. I assumed that we had the investigation of his father under control and that another additional autosomal DNA test or two of strategically chosen cousins would reveal his biological father. My new investigation would involve "stirring the pot" about his biological mother.

I went to GEDmatch.com, where both Jim's results from Ancestry.com and mine from various sources were parked for analysis. I decided to search Jim's xDNA against the database to see if he matched anyone who could be placed on the maternal side of his family. Of course, I came up because we were first cousins. So I concentrated on the others who matched Jim. Only a day later did I realize that I was ignoring the most important of Jim's matches—his match with me. In Figure 5.7, the dark portions of the bar graph at the left (21.2 cM) and right (89.8 cM) ends represent the areas where we had matching segments greater than 3 cM long. They appear to cover about half of our xDNA.

It belatedly dawned on me that if Jim and I share this much of our X chromosomes, our mothers were closely related. Jim's mother must be a sister of my mother. That realization turned our investigation around 180 degrees. Instead of investigating my four uncles to determine which one was Jim's father, we needed to investigate my nine aunts to determine which one was his mother. What a paradigm shift! Had I remembered the rules of xDNA inheritance, I would have saved myself considerable time, money, and resources that were spent in the investigation of my four Adams uncles.

With that realization, another bit of recently acquired information took on new meaning. Jim had recently initiated an effort to find out what information court records might contain. He wrote to me:

> The circuit court clerk was surprised that I was tagged with the family name of the father rather than the mother. Apparently it was more common around here at that time for an unwed mother to pass her family name, not the father's, on to the child.[6]
>
> When I was three years old, in the circuit court I was adopted by Morgan and Florence Jones and my name changed from James Edward Adams to James Edward Jones. The circuit court clerk will be looking at that adoption record and the birth certificate if available. It feels unlikely that there was any earlier name change.[7]

Figure 5.7
xDNA comparisons of the author and his new cousin conducted at GEDmatch.com

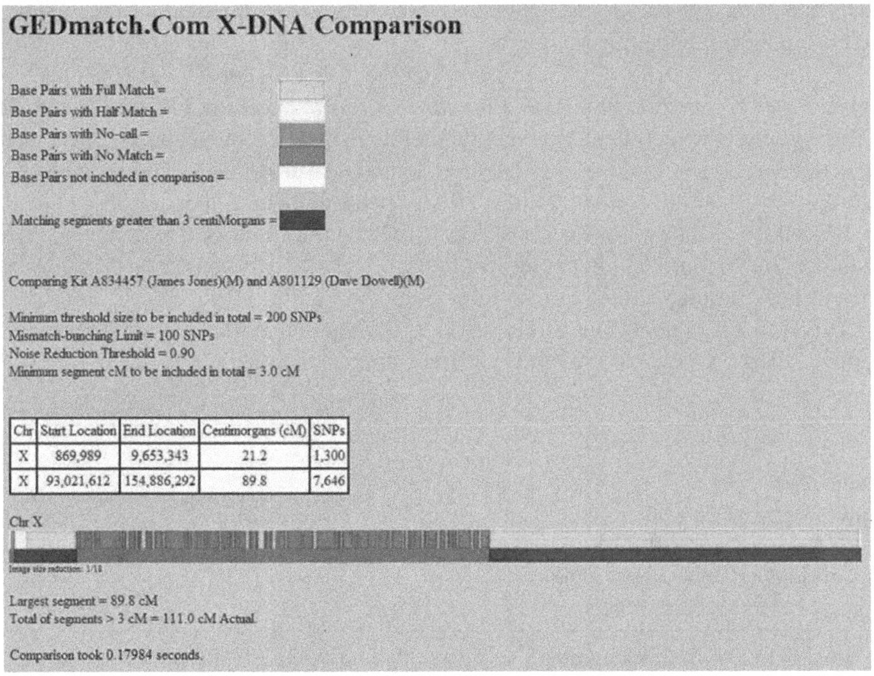

GEDmatch.Com X-DNA Comparison

Base Pairs with Full Match =
Base Pairs with Half Match =
Base Pairs with No-call =
Base Pairs with No Match =
Base Pairs not included in comparison =

Matching segments greater than 3 centiMorgans =

Comparing Kit A834457 (James Jones)(M) and A801129 (Dave Dowell)(M)

Minimum threshold size to be included in total = 200 SNPs
Mismatch-bunching Limit = 100 SNPs
Noise Reduction Threshold = 0.90
Minimum segment cM to be included in total = 3.0 cM

Chr	Start Location	End Location	Centimorgans (cM)	SNPs
X	869,989	9,653,343	21.2	1,300
X	93,021,612	154,886,292	89.8	7,646

Chr X

Image size reduction: 1/11

Largest segment = 89.8 cM
Total of segments > 3 cM = 111.0 cM Actual.

Comparison took 0.17984 seconds.

Based on our xDNA, we now know why the file was tagged the way it was in circuit court: Jim's mother was an Adams. That meant we had a different task. Jim's mother could not be my mother. We share enough DNA to be first cousins. We would need to share about twice as much to be half-siblings. One of my aunts can be eliminated because she gave birth to a daughter the month before Jim was born. Another is a highly unlikely candidate because she gave birth the month Jim was conceived. Of the remaining seven sisters, two others were married and no longer using their Adams maiden name at the critical time. Four were still living at home during the 1930 census, which was recorded about a year and a half before Jim was conceived. These four received our first attention. But if we have learned anything, it is not to overlook any possibility until it has been completely eliminated.

One of the sisters residing at home in 1930 was my Aunt Hattie. From my earliest memories she had been institutionalized in the Missouri State Hopital in St. Joseph. None of my living cousins could recall the circumstances for Aunt Hattie's institutionalization. According to the 1940 census, she was listed as living in the hospital in 1935. This narrowed the window for her hospitalization down to five years which included the time when Jim was conceived. Speculation swirled among my cousins who were working with me to identify Jim's birth mother. Could some kind of traumatic incident have led to her impregnation and caused Hattie to snap and require institutionalization? I petitioned the local court as Hattie's "next of kin" for the release of her hospital records. Thirty pages reproduced from very poor microfilm were subsequently released. Most of her records related to her admission in July 1931 and her death in 1953 from tuberculosis. Her mental affliction was attributed to "scarletina" as scarlet fever was often called at the time. According to the hospital records, this condition had its onset when she was nine years old. Since Hattie had been hospitalized about five months before Jim would have been conceived, she was eliminated as a candidate to be his biological mother.

Such was the status of our search on June 30th, 2014 when I and three other first cousins met Jim for the first time. Five days later Jim received documents from the Circuit Court that identified his birth mother as Esther Yvonne Adams. She was the next to the youngest of the Adams siblings and was 19 at the time Jim was conceived. She married Richard Walker in March, 1935. About six months later and about the time of Jim's third birthday, his name was legally changed to James Edward Jones. Esther later divorce Richard and is not known to have had any children other than Jim. She died in 1997 at age 85.

This search is ongoing and is relying on both traditional research methods and DNA testing tostart the process of identifying Jim's biological father.

Figure 5.8
Summary of xDNA inheritance (Courtesy of Elise Friedman.)

X Chromosome Inheritance Protocols

Female	Male

Women have two X chromosomes.
• One inherited from mother; one inherited from father.
  • X from mother can be inherited from both maternal grandparents.
  • X from fatheralways inherited from the paternal grandmother.
	Men have one X chromosome.
	• Inherited from mother.
	• X always inherited from the man's maternal grandparents.

Daughter inherits X from father. • Dad inherited Y from his father. Therefore, no X from the paternal grandfather or his ancestors. • Father inherited X from his mother. Paternal grandmother inherited one X from each parent.	**Son inherits Y from father.** • No X chromosome from paternal side of family at all.
Daughter inherits X from mother. • Mother inherited X from both parents. • Maternal grandfather inherited Y from his father; therefore no X from the maternal grandfather's paternal side. • Maternal grandfather inherited X from his mother, who inherited X from both parents. • Maternal grandmother inherited X from both parents. Her father inherited one X from his mother. Her mother inherited X from both parents.	**Son inherits X from mother.** • Mother inherited two X chromosomes, one from each parent. So son's X chromosome can be from either maternal grandparent or both. • Maternal grandfather inherited Y from his father, so no X from maternal grandfather's paternal side. Maternal grandfather inherited X from his mother. His mother inherited one X from each of her parents. • Maternal grandmother inherited two X chromosomes, one from each parent. Her father inherited one X from his mother. Her mother inherited from both her parents.

The research techniques for examining xDNA matches are very similar to those followed with atDNA matches. The critical difference is remembering the difference between the two inheritance patterns. Remember—even if your author does not always do so—*boys do not inherit xDNA from their fathers.*

SUMMARY

I will run through this unusual pattern of DNA inheritance one more time with the assistance of Figure 5.8, which contains information extracted from a webinar presented by genetic genealogist and trainer Elise Friedman.[8]

NOTES

1. David R. Dowell, "The Genetic Gods Were Smiling," *Dr. D Digs up Ancestors,* http://blog.ddowell.com, August 12, 2013. http://blog.ddowell.com/2013/08/the-genetic-gods-were-smiling.html, accessed October 6, 2013.

2. James C. Turner, "X Chromosome Inheritance 5 Generation Chart," http://freepages.genealogy.rootsweb.ancestry.com/~hulseberg/DNA/xinheritance.html, accessed October 8, 2013.

3. David R. Dowell, "Another Adams Cousin," *Dr. D Digs up Ancestors,* http://blog.ddowell.com, accessed December 22, 2013.

4. Jim Jones, email to author, November 18, 2013.

5. David R. Dowell, *Crash Course in Genealogy* (Santa Barbara, CA: Libraries Unlimited, 2011), 14, 21–22.

6. Jim Jones, email to author, January 29, 2014.

7. Jim Jones, email to author, January 30, 2014.

8. Content adapted from Elise Friedman, "Family Tree DNA Feature Launch: X-Chromosome Matches in Family Finder," *FTDNA Webinars,* January 7, 2014, http://www.familytreedna.com/learn/ftdna/webinars/, accessed March 1, 2014.

6

What Is Extreme Genealogy?
Haplogroups

"Your story has a surprise beginning."

—Advertisement in *The Atlantic*, September 2012[1]

How far back is the beginning of your story? Traditional family histories trace descendants from some founding ancestor down to the present. In contrast, good research starts with the present and methodically moves back in time. Genetic genealogist and author Debbie Kennett describes the status in which most of the more diligent of us find our family history:

> Most family historians are able to name all their parents, grandparents, great-grandparents and perhaps most of their great-great-grandparents too. With a little bit of luck, and a lot of dedicated research, it is often possible to trace some of those lines several generations further back into the 1600s or even the 1500s. If any of those lines intersect with the aristocracy or royalty the pedigree can usually be taken back several more centuries.[2]

English actress Tilda Swinton, who comes from a family that can be traced back to the ninth century, put it in perspective: "Everyone's from an old family. Mine just wrote everything down."[3]

The DNA analyses described in earlier chapters of this book are extremely helpful in elevating our understanding of who we are and who came before us. However, at some point these tools reach the limits of what they can help us discover about our ancestors. yDNA and mtDNA can give us insight into who some of our ancestors were many, many generations ago. But after the first few generations back, their results reflect only a tiny and decreasing fraction of our ancestors in each generation. While atDNA is more comprehensive, its power to illuminate starts to diminish significantly after about five generations (Figures 6.1 and 6.2).

Blaine Bettinger, who blogs as *The Genetic Genealogist,* describes our "Genealogical Tree, which is every ancestor in history [who] had a child who had a child who had a child [who] ultimately led to you. [Decisions] made by every person in that tree contributed to who and what you are today."[4] Both by nature (DNA) and by nurture

Figure 6.1
Probabilities fade quickly that you will find DNA information from a specific ancestor

Probability of Retrieving Inherited Informative DNA about Ancestors Decreases with Each Passing Generation

| Generation | Percentage of Your Ancestors per Generation about Whom Your DNA Provides Information | | Probability of Identifying Shared DNA |
	From yDNA	From mtDNA	From atDNA
Parents	50%	50%	100%
Grandparents	25%	25%	100%
First great-grandparents	12.5%	12.5%	>99%
Second great-grandparents	6.25%	6.25%	90%
Third great-grandparents	3.125%	3.125%	50%
Fourth great-grandparents	1.56%	1.56%	10–15%
Fifth great-grandparents	0.78%	0.78%	<5%
Sixth great-grandparents	0.39%	0.39%	<2%
Seventh great-grandparents	0.195%	0.195%	<1%
Eighth great-grandparents	0.0977%	0.0977%	<1%

(how those people interacted with their environment), those ancestors have had a small but cumulative role in shaping us. Their daily choices of diet, of whether to move to the next valley, with whom to mate, and so on, have determined whether each of us would be born and where that event would occur. Only in relatively recent generations can we identify specific segments of DNA that we inherited from each of them (see Figure 6.1). Nevertheless, our genomes are defined by the accumulation of SNPs that they sequentially and collectively have passed down.

As shown in Figure 6.1, the probabilities that you will retrieve inherited informative DNA about specific ancestors decrease with each passing generation. By the eighth great-grandparents, that chance is down to 1 percent.

At the truncated top (paternal grandfather) quadrant of the pedigree chart shown in Figure 6.2, note that the ability to "see" the DNA information about ancestors fades quickly except for the "yDNA" line at the top of the chart. The bottom quadrant of this chart, if drawn, would be a mirror image, with only the information from the mtDNA line clearly readable after a few generations. The two middle quadrants fade equally quickly along all branches.

How would you like to be able to extend your family tree back thousands of years? You can begin to do just that and much, much more if you are willing to play by rules that are a little different than those we use in traditional genealogy. I'll call these the *Extreme Genealogy rules*. I could also call them the Anthropological Genealogy rules or the Population Genetics rules, but for my purposes I like Extreme Genealogy rules better. Some may call this genealogical mythology, but it is based on a fast-growing body of scientific research.

Figure 6.2
Most DNA information fades from our view within a few generations

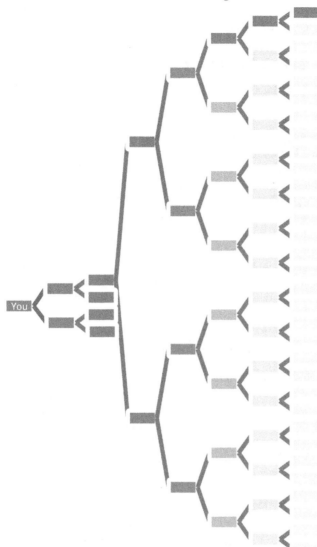

   In this process, we will not be able to uniquely identify everyone in every generation of our trees. What we will be able to do is identify key individuals along at least two of our ancestral lines who have had roles in shaping who we are today. We will not know which names they used in their lifetimes or exactly when and where they lived. However, we can put them in chronological order, approximate where they probably lived, and perhaps deduce something about their language and lifestyle. Spencer Wells, director of the Genographic Project (Geno 2.0) at National Geographic, describes this process:

> So what DNA can tell you is the who, the where, and the when on human migration.
> And then to get at the how and the why, which is the interesting and exciting stuff for

most people, you have to draw in all these other fields. So it's a very multidisciplinary task. You have to draw on archeology, linguistics. You have to draw on, you know, physical anthropology—lots of other aspects. Climatology, it turns out—all sorts of other fields. So, it's a license to be a polymath or a dilettante in a way; working in this field.[5]

SPECIAL RULES FOR EXTREME GENEALOGY

Rule 1: The males and females whom we list in the same generation on our Extreme Pedigree Charts (EPCs) were not mates and probably did not even live at the same time.

Rule 2: There are gaps of unknown periods of time between the generations shown in our EPCs.

   Rule 2a: We will eventually fill in some of these gaps by discovering new SNPs or branching points of our DNA.

   Rule 2b: We will never fill in all the gaps.

Rule 3: The female line of ascent is currently much more complete than that for males. Many of us have had our entire mitochondrion sequenced. We are just beginning to have a few complete Y chromosomes sequenced. While there are *only* 16,659 base pairs along the mitochondria, "the Y chromosome spans more than 59 million building blocks of DNA (base pairs) and represents almost 2 percent of the total DNA in cells."[6] Even though it is one of our smallest chromosomes, tests for the complete Y chromosome are just starting to reach the consumer market. The current price is affordable only to those determined to be trailblazers for the rest of us. However, it is through their contributions that we will all be able to trace in more detail our evolutionary trail back through time.

HAPLOGROUPS

In the simplest definition, a haplogroup is a group of people who all descend from a common ancestor. In genetic genealogy, we attempt to break haplogroups down into smaller and smaller subgroups as we attempt to find the most recent common ancestor who is genetically identifiable as distinctively unique. This ancestor carries a mutation that is inherited by all descendants, thereby defining them as members of the same haplogroup. *The classification schemes for male haplogroups and for female haplogroups are completely separate and independent, despite the fact that both use a system of letters and numbers to describe them.* They go back to either "Y-Chromosome Adam" or "Mitochondrial Eve," respectively.

Y-CHROMOSOME ADAM (yADAM)

The most recent discoveries of SNPs are being driven by citizen scientists as well as by academic labs. One of the most amazing recent discoveries appears to have extended the genetic haplotree of a South Carolina man back to about 338,000 years ago.[7] This certainly qualifies as Extreme Genealogy! This astounding discovery came about not as the result of a carefully thought-out academic research project, but rather as the result of a man taking part in a haplogroup research project managed by volunteer genetic genealogist Bonnie Schrack. Bonnie could find no place to put him in the existing structure of her project. When molecular anthropologist Thomas Krahn, assisted by his wife Astrid, analyzed this swab sample in the FTDNA laboratory, the SNPs just kept on coming. The Krahns unintentionally spent an all-nighter listing all of these mostly new and previously unknown SNPs. Bonnie and the Krahns then collaborated with a multidisciplinary international research team of academics to verify and date their find. By the time they finished, they had extended our human tree back thousands of generations (Mendez, 2013):

> We report the discovery of an African American Y chromosome that carries the ancestral state of all SNPs that defined the basal portion of the Y chromosome phylogenetic tree. We sequenced ~240 kb of this chromosome to identify private, derived mutations on this lineage, which we named A00. We then estimated the time to the most recent common ancestor (TMRCA) for the Y tree as 338 thousand years ago (kya) (95% confidence interval = 237–581 kya). Remarkably, this exceeds current estimates of the mtDNA TMRCA, as well as those of the age of the oldest anatomically modern human fossils. The extremely ancient age combined with the rarity of the A00 lineage, which we also find at very low frequency in central Africa, point to the importance of considering more complex models for the origin of Y chromosome diversity. These models include ancient population structure and the possibility of archaic introgression of Y chromosomes into anatomically modern humans. The A00 lineage was discovered in a large database of consumer samples of African Americans and has not been identified in traditional hunter-gatherer populations from sub-Saharan Africa. This underscores how the stochastic nature of the genealogical process can affect inference from a single locus and warrants caution during the interpretation of the geographic location of divergent branches of the Y chromosome phylogenetic tree for the elucidation of human origins.[8]

This discovery was so groundbreaking that it required that a new haplogroup, A00, be inserted into the human tree trunk before the first previously known group. Bonnie is currently working with a research partner in Africa to collect new swabs that may match the original sample. All of our paternal lines now have a new beginning!

Haplogroup Migrations of yDNA

The map in Figure 6.3 shows the gradual mutations of the Y chromosomes of males over thousands of years. These created genetically distinguishable clans as humans moved out of Africa to populate the Earth. Distinct haplogroups emerged along the migration paths as SNP mutations occurred in the Y chromosomes and cumulated over multiple generations. The letters assigned to these groups have changed as new DNA patterns emerged.

As mentioned earlier, it is important to remember that maternal haplogroups (traced through the mitochondria) and paternal haplogroups (traced through the Y chromosome) are classified by different schemes. Therefore maternal haplogroup classifications cannot be compared with paternal haplogroup classifications. I realize I am being redundant on this point but I cannot state it strongly enough: haplogroup A for women is in no way related to haplogroup A for men. In Figure 6.3, haplogroup A for paternal lines indicates African ancestry. In contrast, haplogroup A for maternal lines indicates Native American heritage. These two schemes might as well have been developed on separate planets. Maybe John Gray had it right when he titled his book *Men Are from Mars, Women Are from Venus*.

In my mind, the male haplogroups classification nomenclature makes more sense than the nomenclature used to describe mitochondrial haplogroups. Major paternal or Y-chromosome haplogroups are designated by a capital letter. The earliest haplogroup known when this scheme was developed was assigned the label haplogroup A.

Figure 6.3
Map showing the migration of Y chromosomes (men) out of Africa to populate the world (Courtesy of www.FamilyTreeDna.com.)

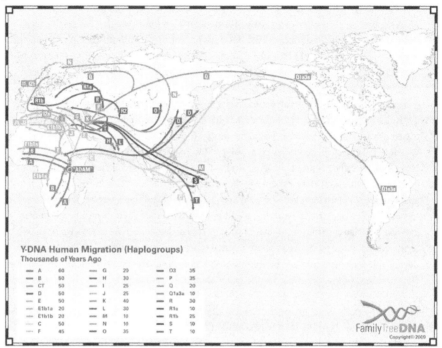

It originated in Africa. After the first major branching mutation was found, the men who carried that SNP forward were classified as belonging to haplogroup B. Then came another major branching, resulting in haplogroup C. In other words, the paternal classification scheme was originally alphabetically based, with the next available letter being assigned after each major branching occurred. Many of the mutations that define smaller and smaller branches have only recently been discovered. Rather than assign a new letter, they are grouped under their original letter with a suffix added. The protocol for designating these subgroups is to follow the original capital letter with a number and then a lowercase letter. As more subdivisions are needed, additional numbers and lowercase letters are appended. Geneticists are discovering new SNPs at an accelerating pace as more of our genomes are explored in detail. As this has happened, the classification scheme, which had a logical and intuitive beginning, has become much more complex—somewhat parallel to what has happened to library classification schemes. For example, one of the most common paternal haplogroups is at this writing now known as R1b1b2a1**b5**—or is it R1b1b2a1**a2f**?

Those of you who are really into classification schemes will note that the first eight characters, reading from left to right, are identical. It is what happens at the extreme right end of this haplogroup designation that seems to be different. Actually, this discrepancy reflects that different labs have more or less simultaneously discovered new subdivisions of the main haplogroup. These two designations are actually two different ways of defining the very same haplogroup. By the time you are reading this book, standardization should have set in and one of these designations likely will have prevailed. My reason for belaboring this point is not to ensure that you remember either one of these designations. Instead, this discussion is meant to remind you of how fast the science of genetics and genetic genealogy is advancing. In some ways, the application of genetics to genealogy is now a process of "Ready, fire, aim!" We cannot stop progress in the lab until we come up with the perfect classification scheme, particularly when we cannot predict what will be discovered next.

A new protocol for expressing yDNA haplotypes is emerging. For a good summary of what this change is and why you may want to use it, read CeCe Moore's blog post on this topic.[9] Under the original system, an alternating new number or letter was tacked on at the end or inserted in the middle of an existing sequence. Under the new system, when a new branching SNP is discovered, the haplogroup terminal SNP alone is used to define that location. "A terminal SNP is the defining SNP of the latest subclade known by current research."[10] It is expressed on its own by using the first letter of the long form and then the youngest/most recently occurring SNP. Thus the two designations given earlier can both be abbreviated as R-L21. This indicates they belong to the major haplogroup R and that the most recent SNP is L21.

The "L" in the L21 SNP is a designation for the lab in which that SNP was discovered.[11] The "21" indicates that it is the 21st SNP discovered in that lab. Because new SNPs are often discovered nearly simultaneously in more than one lab, equivalent SNPs can be assigned a variety of names. For example, R-L21 is listed on the International Society of Genetic Genealogists (ISOGG) tree as equivalent with M529, S145, L459, and Z290. Most SNPs do not have this many alternative names, but this location is very common and defines the largest group of men found along the Atlantic coast of Western Europe.

The current pace at which new SNPs are being discovered is mind boggling. Alice Fairhurst, chair of the ISOGG work group that maintains the Y-tree, reported that the tree had accumulated approximately 2,000 SNPs by the end of 2012 and that an additional 1,400 had been added just between January and June 2013.[12] A couple of major events—the Geno 2.0 project at National Geographic and the Walk through the Y project[13]—distorted the growth curve during this time frame, but new SNPs are still being discovered at rates that would have been unimaginable just a few years ago. Most of these new SNPs are being discovered on the younger side of known defining SNPs; however, a few are much older and are stirring up considerable excitement.

Why should you care about SNPs? Angie Bush explains, "Hopefully if we get enough of them they will become genealogically useful and/or we will gain a better picture of how the human race migrated and evolved over time."[14] It may help to visualize SNPs as permanent exit ramps from the main stream of DNA as it flows down through human history. Every person who descends from the person who exited the main course (i.e., experienced a mutation) at that exit ramp will carry a SNP to document the event. The accumulations of these SNPs provide us with road maps that allow us to trace our ancestors back through prehistory. We are now exploring the possibility of discovering SNPs that will enable us to connect these maps with genealogical time.

Constructing an Extreme Genealogy

Typically, haplogroup analysis is employed to trace only the two extreme lines of one's pedigree chart. The top line in any generation belongs to the paternal (surname providing) male of that generation. His *guY DNA* allows us to trace the migration of his paternal ancestors back to and through prehistory. On the bottom line of that generation of our pedigree charts is the woman whose *umbilical mtDNA* likewise can be traced through her maternal ancestry back to and through prehistory. These two lines offer wonderful opportunities to explore who we are and how we arrived at where we are today. However, these two lines account for only a small fraction of the DNA in our bodies.

Through documentary research, I can trace my paternal surname line with certainty only back to my sixth great-grandfather, who died in 1733. I have been able to reconstruct what his 111 Y-STR markers likely would be if he had had his Y chromosome tested today. I know a lot about the events of his adult life, but I do not know where or exactly when he was born. I also know to which haplogroup he belongs. That latter fact will be the basis for creating his branch of my Extreme Genealogy.

Enterprising genealogists can identify the haplogroups of other lines who have contributed equally to our genomes. To date, I have been able to identify partial haplogroups for eight of my ancestral lines. Six of these are paternal (yDNA) and two are maternal (mtDNA). This was accomplished by judiciously selecting surrogates to test on my behalf and by combing the results pages of published yDNA surname studies. When using the latter method, I make some assumptions that may be less than optimal. Specifically, I identify men who have been tested and who claim as their earliest known male ancestor a man from whom I have traced my descent through documentary research. I then assign their haplogroup to my ancestor. Although this process is not without risk, it is no worse than other genealogical conclusions that build upon the research of others. At worst, it can be accepted as tentatively correct and subject to refinement when additional evidence becomes available.

Testing Family Members as Surrogates

It is quite common for women to recruit a brother, father, male cousin, or nephew to act as a surrogate donor of a Y-chromosome DNA sample for testing to establish paternal lineages. Other clever uses of surrogates can be made to test specific hypotheses. For example, surname DNA projects often test a carefully selected sample of the present-day male descendants of a family patriarch who lived long ago. This can lead to information that is very helpful in grouping his descendants into branches and separating out non-descendants.

Haplotype of Maternal Grandfather and Beyond

I have recruited several cousins to serve as my surrogates to determine the haplotypes for some of my direct-line ancestors when my own DNA would not carry these answers. I have recruited a male first cousin once removed to establish the yDNA paternal haplogroup of my mother's father (Figure 6.4). I have also used the yDNA test results of a male second cousin to determine the paternal haplogroup of my maternal grandmother's father (Figure 6.5).

You will recall from Chapters 2 and 3 that I have ancestors who appear to be early Anabaptist/Mennonite dissidents. The ancestors in this branch of my family tree fled from Switzerland and settled in Pennsylvania. Along their migration path their surname changed from Groff to Graff to Grove. As discussed in Chapter 3, my only exact full

Figure 6.4
Testing a male first cousin once removed to determine the haplogroup of the researcher's maternal grandfather

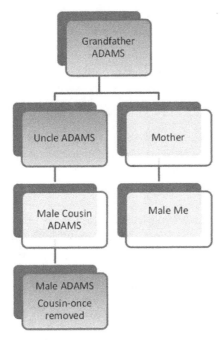

Figure 6.5
Testing a male second cousin to determine the haplogroup of the researcher's great-grandfather

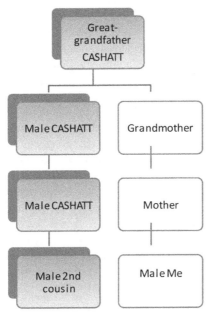

Figure 6.6

Testing a male third cousin once removed to determine the paternal haplogroup of the researcher's great-great-grandfather

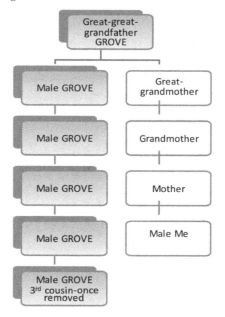

mitochondrial match lives in Switzerland. Gabriela has traced her maternal line back to the middle of the 18th century, all in Switzerland. Both my exact mitochondrial match and I have other ancestral lines we can trace back much further; those lines are early Anabaptist/Mennonite dissidents.[15] My ancestors, whom I can trace through documentary evidence, fled from Switzerland across what is now southern Germany during the 17th century seeking relief from religious persecution. They arrived in Pennsylvania at the dawn of the 18th century.

In addition to my written documentation, I have been able to confirm that my second great-grandfather's yDNA matches that of an eighth cousin three times removed. Our common paternal ancestor was christened in Baretswil, Zurich, Switzerland, in 1616 and died after 1683 in Steinsfurt, Baden, Germany. To make this yDNA comparison, I had to recruit a third cousin once removed to test as my surrogate (Figure 6.6). I told you this is *Extreme Genealogy*: it pushes the limits of what we can know or at least surmise.

I have long collaborated with four other researchers on this Grove line. We are all third cousins of each other. Each of us has extensive databases of our particular trees. When we decided it would be beneficial to use yDNA to verify our paper trail back to Switzerland, we assumed we would have many living male Grove descendants from which to choose someone to test. We were all wrong. None of us had a direct male descendant of our common great-great-grandfather in our databases.

Shortly after we arrived at this conclusion, one of my researcher cousins attended a family reunion in Oklahoma and voila! Three male Groves were in attendance. Two of them did not care about the research and did not want to be bothered. Fortunately, the third was eager to help. Ironically, he lived within 100 miles of me at the time in California.

My cousin surrogate belongs to haplogroup E1b1b1a2. Many of you may already associate haplogroup E with relatively recent African descent. Darvin Martin, a historian of the Anabaptists and a genetic genealogist, explains:

> Nearly all Europeans who test within haplogroup E are part of the E1b1b1a2 subclade. This subclade has direct African origins, representing a recent, common migration from Egypt to Europe within the first few millennia B.C. . . . These Groffs, along with other Mennonite Groff immigrants of Lancaster County, [Pennsylvania] trace their ancestry to Bäretswil, Zürich, Switzerland.[16]

Martin hypothesizes a more distant ancestral migration trail:

> One can reconstruct a theoretical Groff ancestral modal (the STR values of Heinrich Graf of 1356, for example).
> Prior to their arrival in Switzerland the Groffs likely sojourned in modern-day Slovenia. This migratory pattern is further advanced as one sets up a concentration profile of the E1b1b1a2 haplogroup and finds its center in historic Macedonia.
> Before the ancestors of the Groffs entered Switzerland, they lived in Macedonia. Slovenia is on the trajectory between Macedonia and Bäretswil. The fact that Groffs share a common ancestry with some Slovenians is evident as one traces the ancestral migration pattern of haplogroup E1b1b1a2.
> DNA cannot reveal the names of our ancestors, but it can reveal the general locations where our ancestors lived and the approximate times they lived in each location.
> Historically speaking, it is probable that the Groffs were part of a Greek or Roman expansion into points north of the Alps, occurring at the earliest about 500 B.C.
> They were frontiersmen, pushing the boundaries of Roman territory into uncharted areas north.[17]

The E1b1b1a2 haplogroup designation is the long form of the same shortened form, given as E-V13 or even V13.

This part of my family tree is particularly interesting because at least some of the ancestors from whom it is comprised followed a different migration pattern into Europe than was followed by many Europeans. While many of the *Homo sapiens* from whom today's Europeans descend are thought to have arrived on the European continent prior to the last Ice Age, this group appears to have arrived only after the glaciers were in retreat. For most of them, their migration pattern hugged the Mediterranean coast (see the migration path for the E1b1b1 group in Figure 6.3) and did not go as far into Central Asia as had earlier arrivals (see the migration path for the R1b group in Figure 6.3). Some of them may have arrived in Italy by sea from North Africa. *Wikipedia* describes the E1b1b1a2 group:

> E-V13 is in any case often described in population genetics as one of the components of the European genetic composition which shows a relatively recent link of populations from the Middle East, entering Europe and presumably associated with bringing new technologies. As such, it is also sometimes remarked that it is a relatively recent genetic movement out of Africa into Eurasia, and has been described as "a signal for a separate late-Pleistocene migration from Africa to Europe over the Sinai."
> After its initial entry in Europe, there was then a dispersal from the Balkans into the rest of Europe . . . [that] seems to have mainly followed the river waterways connecting

the southern Balkans to north-central Europe." Bird (2007) proposes a still more recent dispersal out of the Balkans, around the time of the Roman empire [sic.].[18]

MITOCHONDRIAL EVE (MTEVE)

Recall from Chapter 3 that mtEve is the earliest woman who has passed mtDNA down to currently living humans along at least two separate lines of descent. Until 2011, it was thought that mtEve lived considerably earlier than yAdam. At the time, the generally accepted science pointed to yAdam living approximately 60,000 years ago. By comparison, mtEve was thought to have lived approximately 140,000 years ago. As we have seen, more recent discoveries have since moved yAdam's era more than twice as far back as that of mtEve. As you probably have surmised, these conclusions are far from settled science; rather, they are just the best projections we can make based on what we have discovered so far.

Haplotype Migration of mtDNA

Also recall from Chapter 3 that the mutations of our mtDNA, as they gradually accumulate, allow us to distinguish between different paths of human migration. The finite size of our mitochondria enables us to get a more complete overview of the stages of these migration patterns than do the incompletely mapped and much more complex yDNA.

Of course, in some cases, females followed parallel migration paths to those of the men. In other cases, men traveled without females and found mates in the course of their travels. It is important to remember that the classification scheme that has been developed to describe the evolution of SNPs on the mitochondria, as it gradually mutated over the centuries, uses letters in a different pattern than the scheme used to describe a similar mutation and migration pattern for men.[19] The mutations that led to distinctly new groupings occurred independently in women (Figure 6.7) and in men (Figure 6.3). Women passed these SNPs along to all of their offspring, whereas men passed them on only to their sons.

Notice that the letter symbolizing the haplogroup for mtEve is not "A," as it was for yAdam. Instead, the mitochondrial nomenclature starts out with the letter "L." Haplogroup A for mtDNA can be found in Siberia and the Americas. Any similarities in other haplogroup designations between mtDNA and yDNA are only coincidental and are confused at your peril.

I can trace my own maternal "umbilical" line with certainty only back to my second great-grandmother. She lived from 1821 to 1910. I believe her mother (or grandmother) was named Elizabeth and was born about 1775 in Pennsylvania. My maternal haplogroup is H13a1a1, which she (and Elizabeth) shares because I inherited it from her. I have already discussed my single exact full mitochondrial match with Gabriela (and her son and daughter) in Switzerland. To obtain information on other maternal lines, I had to find strategically located close cousins to test.

In like manner to the male cousins whose testing was described earlier in this chapter, surrogate donors can be useful for mitochondrial testing. As discussed in Chapter 3, my paper research trail indicated that a sixth great-grandfather, Henry Stedham, had taken as his third wife a woman reportedly named Marjory Oins. Marjory was a direct maternal ancestress of my grandmother. Unfortunately, it was the wrong grandmother.

Figure 6.7

Map of the mitochondrial haplogroup (female) migration out of Africa to populate the world (Courtesy of www.FamilyTreeDna.com.)

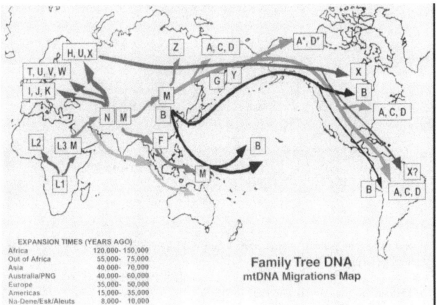

I could not use my mitochondrial DNA sample to test for that relationship because Marjory was an ancestress of my father through his mother. My father could not pass on his mother's mitochondrial DNA to me because my mitochondrial DNA came entirely from my mother. As emphasized previously, only mothers can provide mitochondrial DNA to their descendants. In addition, my father was deceased by the time I identified this research question. To explore this relationship I recruited a first cousin who is the daughter of my father's sister.

Figure 6.8 shows the result when a female cousin gave a surrogate mitochondrial DNA sample to help a male test his paternal grandmother's line. My cousin is a descendant back to Marjory in an unbroken female line of eight generations (two of which are shown in Figure 6.8). As I mentioned earlier, I was fortunate to have my cousin exactly match individuals in the database of Family Tree DNA (FTDNA) who live in Finland and others believed to be of Finnish American descent. The DNA was talking to me and suggesting an answer to my genealogical question. That answer was compatible with a possibility that the documentary evidence suggested.

DEEP ANCESTRY OR EXTREME GENEALOGY?

By combining traditional research with targeted DNA testing, I have learned some extreme things about my family that I had previously been unable to unravel. Some of them include the following findings:

• Many of my male lines (six lines to date and counting) entered Europe carrying a R1b1a2 (or R-M269) SNP as part of the mass migration from Western Asia (see Figures 6.9 and 6.10).

Figure 6.8

Testing a female first cousin to discover the researcher's paternal grandmother's haplogroup

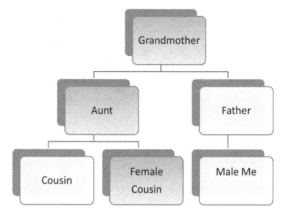

- At least two of my male lines who carried the E-V13 SNP entered Europe by taking a more Mediterranean path.
- A great-great-grandmother whom I can trace back only to Pennsylvania around 1800 probably had an Anabaptist ancestress who lived in Switzerland about 1600.
- A sixth great-grandmother whom I can trace back only to 1736 in Delaware had an ancestress who lived in Finland around 1600.

These are but a few examples of the many genealogical hypotheses that can be tested by carefully choosing the right relative or relatives and then persuading them to swab their cheeks or spit into sample tubes. I hope these examples will stimulate your imagination in ways that will allow you to explore wider areas of your pedigree chart.

The six-generation pedigree chart in Figure 6.9 shows haplogroups of eight ancestral lines that would extend this tree back into history for thousands of years. The haplogroup membership in Figure 6.10 shows eight of the author's 16 great-great grandparents from the pedigree chart in Figure 6.9.

A word of caution needs to be inserted here. In Figure 6.10, the haplogroups of Jacob Brown and James Pierce are both shown as R1b1a2. They would appear to be closer to each other than to other ancestors. While this possibly could be true given that each of their ancestral lines can be traced back to England in the 16th century, the information presented here is somewhat misleading. Both are reported to be R1b1a2 in Figure 6.10 because my surrogates have not yet tested their haplogroups more completely. If they were to do so, it is probable that their haplogroup designation would be extended to one of the subclades that contains even more letters and numbers. However, their "backbone" root stub of R1b1a2 would still be held in common. R1b1a2, which is now becoming more widely known as R-M269 for its terminal defining SNP, is the most common clade found in Western Europe. *Wikipedia* reports its "frequency is about 71% in Scotland, 70% in Spain and 60% in France. In south-eastern England the frequency of this clade is about 70%; in parts of the rest of north and western England, Spain, Portugal, Wales and Ireland, it is as high as 90%; and in parts of north-western Ireland it reaches 98%."[20] With this kind of penetration along the Atlantic coast

Figure 6.9

Pedigree chart showing haplogroups of great-great-grandparents

Pedigree Chart - David Ray DOWELL Ph.D.   3 December 2013

Chart no. _____
No. 1 on this chart is the same as no. _____ on chart no. _____

1 David Ray DOWELL Ph.D.
b:
p:
m:
p:
d:
p:
sp:

2 Clarence Ray DOWELL
b: March 15, 1913
p: Bristol, Prowers, CO
m: April 19, 1940
p: Trenton, Grundy, MO
d: August 2, 2006
p: Chillicothe, Livingston, MO

3 Lucille Ruth ADAMS
b: January 23, 1907
p: Ludlow, Livingston, MO
d: August 1, 1998
p: Chillicothe, Livingston, MO

4 Arthur Ernest DOWELL
b: February 26, 1882
p: Morehead, Neosho, KS
m: September 1, 1909
p: Holly, Prowers, CO
d: February 11, 1937
p: Bakersfield, Kern, CA

5 Mary Jane PIERCE
b: August 18, 1890
p: Wheeler, Charles Mix, SD
d: May 26, 1927
p: Denver, Denver, CO

6 Cash Clay ADAMS
b: April 6, 1870
p: Frankford, Pike, MO
m: December 26, 1893
p: , Daviess, MO
d: December 7, 1955
p: Breckenridge, Caldwell, MO

7 Anna Lois CASHATT
b: May 4, 1876
p: Cincinnati, , OH
d: October 22, 1943
p: Breckenridge, Caldwell, MO

8 Lindsia E. DOWELL
b: December 22, 1849
p: Gilman City, Harrison, MO
m: 1879
p: , , KS
d: February 14, 1930
p: Morehead, Neosho, KS

9 Adaline Rebecca MOORE
b: November, 1857
p: Brodhead, Green, WI
d: April 15, 1935
p: Morehead, Neosho, KS

10 James A. PIERCE Jr.
b: February 20, 1860
p: , Bureau, IL
m: 1883
p:
d: February 8, 1929
p: Holly, Prowers, CO

11 Emma Mary HOAR
b: November 19, 1866
p: Boaz, Richland, WI
d: September 26, 1941
p: Holly, Prowers, CO

12 Benjamin Franklin ADAMS
b: about 1837
p: , , KY
m: May 5, 1864
p: , Pike, MO
d: about 1875
p: Maryville, Nodaway, MO

13 Margaret Ann MOORE
b: about 1843
p: , , IN
d:
p:

14 Clark CASHATT
b: March 18, 1838
p: , Highland, OH
m: June 22, 1869
p:
d: September 22, 1907
p: Maryville, Nodaway, MO

15 Sarah Americas GROVE
b: November 15, 1845
p: Hebron, Licking, OH
d: August 15, 1936
p: Breckenridge, Caldwell, MO

16 Peter Clingman DOWELL
b: July 24, 1826
d: March 18, 1878

17 Hester Delaney BROWN
b: January 10, 1827
d: 1863

18 Andrew Jackson MOORE
b: June 5, 1828
d: June 12, 1894

19 Nancy Shedd Richardson
b: November 15, 1832
d: 1899

20 James Arvin PIERCE Sr.
b: May 20, 1821
d: January 7, 1892

21 Mary Jane PERRY
b: January 1, 1834
d: January 9, 1905

22 John David HOAR
b: December 2, 1826
d: November 24, 1902

23 Hannah Maria PYLE
b: June 6, 1833
d: March 21, 1904

24 Isaac N. ADAMS
b: about 1813
d: after 1870

25 Patsey Reeves
b: about 1813
d: before 1870

26 William MOORE
b:
d: before 1850

27 Sarah
b: about 1804
d:

28 Daniel Faulkner CASHATT
b: August, 1815
d: 1883

29 Amy C. HALL
b: about 1816
d: December 4, 1874

30 Samuel GROVE Jr.
b: November 1, 1818
d: May 18, 1899

31 Mary Ann SHOVER
b: November 18, 1821
d: May 2, 1910

32 R1b1a2a1a1b4 DOWELL
33
34 R1b1a2 BROWN
35
36
37
38
39
40 R1b1a2 PIERCE
41
42
43
44
45
46
47 U5b1b2 LODGE
48 R1b1a2a1a1b5a ADAMS
49
50
51
52
53
54
55
56 R1b1a2a1a1b CASHATT
57
58
59
60 E1b1b1a2 GROVE
61
62
63 H13a1a1

Prepared 3 December 2013 by:
David R. Dowell
http://blog.ddowell.com

of Europe, it is not unexpected that the paternal haplogroups of five of my six great-grandfathers belong to the super-clan R1b1a2.

The remaining male haplogroup, E1b1b1a2 (E-V13), represents one of my lines that, according to the paper trail, never got close to northern Europe as it migrated toward North America. However, this haplogroup is found in northern Europe, even within my family tree. I have a ninth great-grandfather who turns up in southern Sweden early in the 17th century. He was thought to have been a traveling salesman

Figure 6.10
Haplogroup membership of eight of the author's 16 great-grandparents

Ancient Ethnic Tribes

Great-Great Grandparent	Haplogroup
Peter Clingman Dowell	R1b1a2a1a1b4
Jacob Brown	R1b1a2
James Arvin Pierce	R1b1a2
Mary Jane (Perry) Pierce	U5b1b2
Isaac N. Adams	R1b1a2a1a1b5a
Daniel Faulkner Cashatt	R1b1a2a1a1b
Samuel Grove, Jr.	E1b1b1a2
Mary Ann (Shover) Grove	H13a1a1

from Denmark, but I have found no documentation to confirm this point of origin. A living descendant has been confirmed to carry the E-V13 SNP. Researchers have found traces of that haplogroup along the rivers flowing northward from the Alps toward the North Sea. Some have theorized that this migration was facilitated by the Romans as they moved toward Britain almost two millennia ago.

PEDIGREE COLLAPSE

If you are an experienced genealogist, you will realize that pedigree charts have limits in how far they can expand geometrically. We do not have to go back very many generations before there would not have been enough people living on our planet to fill all the boxes on our charts without some of the individuals occupying more than one box. Cecil Adams (possibly a pseudonym), writing in "The Straight Dope," a popular question-and-answer column published in *The Chicago Reader* for the last four · decades, says:

> If you go back far enough, however, pedigree collapse happens to everybody. Think of your personal family tree as a diamond-shaped array imposed on the ever-spreading fan of human generations. (I told you this was cosmic.) As you trace your pedigree back, the number of ancestors in each generation increases steadily up to a point, then slows, stops, and finally collapses. Go back far enough and no doubt you would find that you and all your ancestors were descended from the first human tribe in some remote Mesopotamian village. Or, if you like, from Adam and Eve in the Garden of Eden.[21]

Many of us have already documented cousin marriages that cause redundancy in the ancestors we have listed in our charts. Richard Cunndiff, writing in *Discover*, explained it in this way:

Until the past century, families tended to remain in the same area for generations, and men typically went courting no more than about five miles from home—the distance they could walk out and back on their day off from work. As a result, according to Robin Fox, a professor of anthropology at Rutgers University, it's likely that 80 percent of all marriages in history have been between second cousins or closer.[22]

You may be surprised to find that first cousin marriage is still legal in several U.S. states.[23]

Population bottlenecks have occurred at various times that have greatly reduced the choices of available mates. We do not know much about the earliest bottlenecks, which may have occurred in Africa more than 50,000 years ago. However, for those of us who have European ancestors, our predecessors moved out to Asia in relatively small bands. These groups did not provide diverse choices from which to find mates. Some of these groups became extinct and, therefore, left no present-day descendants. Others hovered on the edge of extinction at various times.

Changes in climate played a key role. Although we believe that some of our ancestors had made their way into Europe prior to 15,000 years ago, the expansion of the polar ice masses into lower latitudes forced them to move close to the Mediterranean and adopt other survival strategies as they faced challenges that reduced the population. As the glaciers expanded, they trapped more and more of the water that would have been part of the oceans, which lowered the ocean levels by as much as 300 feet and allowed our ancestors to move from what is now one landmass to another over land bridges between Siberia and North America, the British Isles and continental Europe, and possibly even Southeast Asia and Australia. These ice ages coincided with severe changes in the climate of the Sahara as well as other parts of the Earth.

More recently, religious and tribal taboos against marrying outside the group, wars, and epidemics have created population bottlenecks by reducing the number of available or acceptable partners. These and other events beyond the scope of this book all have contributed to pedigree collapse—the reduction of the total number of individuals from whom we descend.

At which generation in our Extreme Genealogy does the number of our discrete ancestors peak? At least one model predicts this might have been around 1200 AD.[24] This point would vary significantly from one geographical and cultural group to another. In many cultures, large numbers of present-day individuals appear to descend from a relatively few founding individuals who lived in the last 1,000 or 2,000 years.

If I were to create a pedigree chart to trace my ancestry back in time, my paternal lines would definitely collapse by the time I reached the first male to carry the R1b1a2 (or R-M269) SNP mutation. By then, many—if not most—of my paternal lines would have merged. Five of the six paternal lines for which I know the haplogroup memberships for my great-great-grandfathers share that SNP (see Figures 6.9 and 6.10). I do not have a clear picture of whether my maternal lines would be that homogeneous, but I doubt it. I would expect more diversity in that part of my pedigree chart.

ROYAL HAPLOGROUPS

Back at the beginning of this chapter, Debbie Kennett suggested, "If any of [our] lines intersect with the aristocracy or royalty the pedigree can usually be taken back several more centuries."[25] If you want to have some fun fanaticizing about such possibilities, you can check the known haplogroup memberships of some royals at "Haplogroups of European kings and queens" (http://www.eupedia.com/forum/threads/25236-Haplogroups-of-European-kings-and-queens).

FOLLOWING YOUR DEEP ANCESTRY

If you have tested at 23andMe or FTDNA, each of these companies provides some tools to help you construct your extreme family tree.

23andMe

1. Click the "My Results" tab.
2. Select Ancestral Tools.
3. Select Haplogroup Tree Mutations Mapper.
4. Select either "Paternal" or "Maternal," and click on your haplogroup, which will be displayed below the box.
5. All defining mutations will be listed from the most recent back to the beginning.

You can then trace the development of your haplogroup from the beginning down to living individuals.

FTDNA (yDNA)

1. For paternal lineage, select "yDNA Haplotree & SNPs."
2. Trace your tree from "Adam."
3. Click "Your Match" at the right side of the screen to extend your tree down to your personal result. Each defining SNP is listed in green (if you have been tested for that SNP) as they branch off to define your tree.

FTDNA (mtDNA)

1. Register to create a "My mtDNA" Community account at http://www.mtdnacommunity.org/ if you decide to proceed.
   a. "The mtDNA Community (http://www.mtDNACommunity.org) pages are a free public service. They are hosted by Family Tree DNA. The site's goal is to help with what we know about the human maternal tree. It is run by scientists outside Family Tree DNA."[26]
   b. Walking the tightrope between confidentiality and access FTDNA hosts this service although it neither does not recommend or endorse it. More about this type of dilemma will be discussed in Chapter 7.
   c. Although mtDNA Community site does provide some useful tools for analysis, be sure you read and understand the "**Privacy Alert:** By uploading mtDNA full sequence results to mtDNA Community, you are making them public."[27] Bloger Roberta Estes elaborates, "If you authorize your full sequence results to be uploaded to mtDNA Community you are authorizing your results to be included in scientific research. In the mtDNA Community, you are not anonymous. This means that your sequence can be tracked back to you. This is neither a bad thing or a good thing, it's just the way it works."[28]
2. From the "Home" tab, click "Upload My Results" at the lower-right corner of the screen.
3. Mouse over "Resources" and click on "Phylogeny."

4. Move the slider down the right side of the screen until you see your macro haplogroup (e.g., select H13 if your haplogroup is H13a1a1).
5. Click on the + symbol to the right of your group to expand the listings.
6. Continue to click on the + symbols that open in indented subgroups until you see your complete haplogroup displayed.
7. Click on your fully expanded haplogroup (e.g., H13a1a1) for more information and analysis.

Geno 2.0

You may choose either the "Your Story" tab or the "Your Map" tab to lead you to the information about the major SNPs in your ancestors' journey. If you are male, choose the paternal or maternal paths you wish to explore. Females are limited to their maternal paths.

AncestryDNA

At this writing, AncestryDNA had not released similar tools to aid in analysis, but the company is said to have some tools under development.

DNA AS A GPS?

Most genetic genealogists are seeking a way to use the information in our genome as a sort of global positioning system (GPS) to point us to the location of our ancestors in the early part of the genealogical era—say, the 17th or 18th century. Will this be possible? The emerging study of biogeography is claiming it will move us in that direction by giving us a new GPS: geographic population structure.

> The search for a method that utilizes biological information to predict human's place of origin has occupied scientists for millennia. Modern biogeography methods are accurate to 700 km in Europe but are highly inaccurate elsewhere, particularly in Southeast Asia and Oceania.
>
> [W]e developed an admixture-based Geographic Population Structure (GPS) method that infers the biogeography of worldwide individuals down to their village of origin.
>
> GPS correctly placed 80% of worldwide individuals within their country of origin. . . . Applied to over 200 Sardinians villagers of both sexes, GPS placed a quarter of them within their villages and most of the remaining within 50 km of their villages, allowing us to identify the demographic processes that shaped the Sardinian society.
>
> Our findings demonstrate the potential of the GenoChip array for genetic anthropology.[29]

While this is a noble goal, the biogeographical version of GPS, as described here, is not yet ready for prime time, even though it may help us a little with our Extreme Genealogies. Placing our ancestors within 700 km in Europe and even less specifically in Asia, while helpful, leaves much to be desired.

The use of haplogroups to try to match individuals to each other is also not ready for prime time. Talking specifically about mitochondrial matching, genetic genealogist and technologist James Lick describes our current status:

> Almost everyone is an imperfect match. This is especially true with incomplete tests like 23andMe but even with full sequences it is rare for someone to EXACTLY match

their haplogroup. The better measure is how well you match the haplogroup and with 3 matching markers for the last hop in the path; that's pretty decent.

Extra mutations are those markers you have in addition to your best haplogroup match. They may form new branches in the future if more matching sequences are found.[30]

Summary

Extreme Genealogy is what I call this research into deep ancestry. I cannot call it X genealogy because that has been pre-empted by chromosome X and its special characteristics, as discussed in Chapter 5. The fun part of Extreme Genealogy is exploring how your paternal and maternal lines got from their beginnings to where in the world you are today. By tracing haplogroups, genetic genealogists can extend their pedigree charts back thousands of generations. In so doing, we can learn a lot about the incredible journeys of groups of our ancestors, even though the precise details of most of our individual forbearers are still unknown and probably forever will remain lost to history. But then how many of us know even a little bit about the daily lives of our 16 great-great-grandparents? In the 1880s, they were about the age we are now. Many of us do not even know all their names. In that regard, Extreme Genealogy is not so different than traditional genealogy.

NOTES

1. National Geographic, ad for The Genographic Project (Geno 2.0), *The Atlantic* (September 2012).

2. Debbie Kennett, *DNA and Social Networking: A Guide to Genealogy in the Twenty-First Century* (Stroud, UK: History Press, 2011), 52.

3. Tilda Swinton, in Buzzy Jackson, *Shaking the Family Tree: Blue Bloods, Black Sheep, and Other Obsessions of an Accidental Genealogist* (New York, NY: Touchstone, 2010), 17.

4. Blaine Bettinger, "Q&A: Everyone Has Two Family Trees—A Genealogical Tree and a Genetic Tree," *The Genetic Genealogist,* posted November 10, 2009.

5. Grant Brunner, "Discussing the Personal Genomics Revolution with Nat Geo's Dr. Spencer Wells," *ExtremeTech* (November 7, 2013), http://www.extremetech.com/extreme/168586 -discussing-the-personal-genomics-revolution-with-nat-geos-dr-spencer-wells, accessed November 8, 2013.

6. "Y Chromosome," in U.S. National Library of Medicine (NLM), National Institutes of Health (NIH), *Genetics Home Reference: Your Guide to Understanding Conditions,* http:// ghr.nlm.nih.gov/chromosome/Y, accessed August 7, 2013.

7. Fernando L. Mendez et al., "An African American Paternal Lineage Adds an Extremely Ancient Root to the Human Y Chromosome Phylogenetic Tree," *American Journal of Human Genetics* 92, 454–459; March 7, 2013).

8. Mendez et al., "An African American Paternal Lineage."

9. CeCe Moore, "Let's All Start Using Terminal SNP Labels instead of Y Haplogroup Subclade Names, Okay?," *Your Genetic Genealogist* (September 20, 2012), http://www .yourgeneticgenealogist.com/2012/09/lets-all-start-using-terminal-snp.html, accessed March 25, 2014.

10. ISOGG, "Y-DNA Project Help: SNP." http://www.isogg.org/wiki/Y-DNA_project_help, accessed March 25, 2014.

11. In this instance, the "L" designates the Family Tree DNA Genomic Research Center in Houston, Texas.

12. Alice Fairhurst, presentation at Genealogy Jamboree, Burbank, CA, June 2013.

13. "Walk through the Y," *ISOGG Wiki,* http://www.isogg.org/wiki/Walk_Through_the_Y, accessed October 24, 2013.

14. Angie Bush, communication with the author, March 24, 2014.

15. Darvin L. Martin, "Unveiling the Deep Ancestry of Swiss Anabaptist Forebears," *Pennsylvania Mennonite Ancestry* 33, no. 3 (July 2010): 2.

16. Martin, "Unveiling the Deep Ancestry," 7.

17. Martin, "Unveiling the Deep Ancestry," 8.

18. "E1b1b1a2 (V13) Early Migration from the Middle East to Europe," April 5, 2011, http://www.e1b1b1-m35.info/2011/04/e1b1b1a2-v13-early-migration-from.html, accessed October 11, 2013. For a fuller discussion of these migrations, see "Haplogroup E-V68," *Wikipedia,* http://en.wikipedia.org/wiki/Haplogroup_E1b1b1a_(Y-DNA)#Early_migration_from_the_Middle_East_to_Europe, accessed November 7, 2013. I am aware that wikis are considered by some to be less than reliable. However, in areas at the very edge of scientific exploration, we are not dealing with "settled science." In such cases, wikis may be our only sources of early information, even though their content may need to be updated as new discoveries are made. In the case described here, this *Wikipedia* article gives a theory that explains the migration of two branches of my family. Of course, that theory will need to be reviewed as we learn more.

19. A quite readable, informative, and widely available book on the arrival of different mitochondrial haplogroups in Europe is the 2001 best seller by Bryan Sykes, *The Seven Daughters of Eve.*

20. "R1b1a2 (R-M269)," *Wikipedia*, http://en.wikipedia.org/wiki/Haplogroup_R1b_%28Y-DNA%29#R1b1a2_.28R-M269.29, accessed November 8, 2013.

21. Cecil Adams, "2, 4, 8, 16 . . . How Can You Always Have MORE Ancestors as You Go Back in Time?", *A Straight Dope Classic from Cecil's Storehouse of Human Knowledge,* August 21, 1987, http://www.straightdope.com/columns/read/412/2-4-8-16-how-can-you-always-have-more-ancestors-as-you-go-back-in-time, accessed March 16, 2014.

22. Richard Conniff, "Go ahead, Kiss Your Cousin: Heck, Marry Her If You Want to," *Discovery* (August 1, 2003), http://discovermagazine.com/2003/aug/featkiss#.UyXCSvldXh5, accessed March 16, 2014.

23. "Cousin Marriage Law in the United States by State," *Wikipedia,* http://en.wikipedia.org/wiki/Cousin_marriage_law_in_the_United_States_by_state, accessed March 16, 2014.

24. Tim Urban, "Your Family: Past, Present, and Future," *Wait But Why,* January 2014, http://waitbutwhy.com/2014/01/your-family-past-present-and-future.html?utm_content=buffer1abb0&utm_medium=social&utm_source=twitter.com&utm_campaign=buffer, accessed March 15, 2014.

25. Kennett, *DNA and Social Networking.*

26. Family Tree DNA, "What is mtDNA Community?https://www.familytreedna.com/learn/scientific-collaboration/mtdna-community/mtdna-community/, accessed August 2, 2014.

27. FTDNA, "Do you recommend that I take part in mtDNA Community?"https://www.familytreedna.com/learn/scientific-collaboration/mtdna-community/recommend-taking-part/, accessed August 2, 2014.

28. Roberta Estes, "The mtDNA Community," *DNAeXplained – Genetic Genealogy Discovering Your Ancestors – One Gene at a Time.* Posted on July 16, 2012, http://dna-explained.com/2012/07/16/the-mtdna-community/, accessed August 2, 2014.

29. E. Elhaik et al., "Geographic Population Structure (GPS) of Worldwide Human Populations Infers Biogeographical Origin down to Home Village," abstract presented at the 63rd

Annual Meeting of the American Society of Human Genetics, http://www.ashg.org/2013meeting/abstracts/fulltext/f130120092.htm, accessed October 25, 2013.

30. James Lick, post to International Society of Genetic Genealogy (ISOGG), *Facebook*, October 21, 2013, https://www.facebook.com/groups/11416337921/10151964064907922/?notif_t=group_activity, accessed October 22, 2013.

7

Is It Ethical? Balancing Technological Possibilities with Human Values

"It is appallingly obvious that our technology exceeds our humanity."

—Albert Einstein[1]

"Changes in the way we way we produce and interact with information lead to changes in the rules we use to govern ourselves, and in the values society needs to protect."

—Viktor Mayer Schonberger and Kenneth Cukier[2]

"Everybody gets so much information all day long that they lose their common sense."

—Gertrude Stein[3]

The birth of "big data" was chronicled by Schonberger and Cukier in their 2013 best seller *Big Data: A Revolution That Will Transform How We Live, Work, and Think*. They wrote, "The sciences like astronomy and genomics, which first experienced the explosion in the 2000s, coined the term 'big data.' "[4] Einstein and Stein also could have been foretelling how "big data" would facilitate DNA testing. The question is shifting from *can* we explore the information in our genome to *should* we? We certainly do not want to diminish our *humanity* or to lose our *common sense* in our rush to explore the inner us.

Throughout history, there has been a struggle to find the appropriate balance between technology and human values. When did this dynamic begin? Was it when the first caveman discovered he could be a more productive hunter if he used a club? How should this new discovery be used? Was it acceptable to use it to keep rivals away? Was it acceptable to use the club to control others? These questions are still relevant to our latest technological advances.

This phenomenon of a technology developing much faster than our understanding of its consequences is far from unique to DNA testing. One more recent example than the club of the caveman is the invention of dynamite. Alfred Nobel's original motivation was to harness the power of nitroglycerine so that it could be used safely in construction work. The fortune he amassed from this and related inventions allowed Nobel to create posthumously some of our most coveted prizes for scientific and humane pursuits. Unfortunately, the new technology had an unintended consequence of igniting an arms race in the last third of the 19th century. All technologies develop faster than do the social policies to guide their appropriate use. They often have far reaching consequences both intended and unimagined. DNA testing is no different.

EXISTING NORMS FOR GENEALOGICAL RESEARCH

Genealogists have long had codes of ethics. While these have been important, they generally have not been uniquely separated from guidelines that govern behavior in other kinds of civilized activity. Genetic genealogist and attorney Judy Russell boils them down to three simple principles:

1. Tell the truth.
2. Play nice with others.
3. Don't tell tales out of school.[5]

Most of the time these guidelines have been sufficient for most of us in our family history research in the 20th century.

More elaborate codes have been developed for those who practice genealogy as a profession. The elements added by these ethics documents are aimed largely at obligations to advance the practice of genealogy and to be professional in the treatment of clients.[6] While ethics can be considered to be professional standards of conduct, many of us who focus almost exclusively on researching our own families may fail to see the relevance of these concepts to our own research activities.

It is not my purpose in writing this chapter to set forth answers to the plethora of moral dilemmas that present themselves as we struggle to use the information from DNA for the benefit of humankind both individually and collectively. Rather, my purpose is to sketch out some of the conflicting concerns that we must balance to arrive at the best answers we can reach to guide our actions in any given situation. If there were obvious and universal answers, this chapter would be superfluous.

BEHAVIOR IN CYBERSPACE

Students in my course "Ethics in the Information Age" were asked to decide whether standards of ethical behavior have changed now that many of our actions are performed in the semi-anonymity of cyberspace. After much deliberation, most of them concluded that ethical standards have not changed but that the situations in which we must apply them change rapidly. While information was previously considered to be found in fixed locations, it now sometimes appeared to be more fluid.

DNA "EXCEPTIONALISM"

Exceptionalism is a very value-laden term. It has been used to characterize the claims of many to be treated differently. Claims of any ethnic, religious, or political group to be considered somehow "chosen" divide us as humans. Our DNA, taken as a whole, proves that we overwhelmingly are the same. Our differences at the genome level are extremely minute.

Blaine Bettinger, in a post in his blog *The Genetic Genealogist*, challenged the assumption that we need new standards for handling the information from our genomes. He was responding to an editorial in the *National Genealogical Society Quarterly* in which genealogists were called upon to establish standards for handling information derived from our genomes.[7] Bettinger said of the editorial:

> [It] operates on an assumption of "DNA exceptionalism," the belief that genetic information is special and should be treated differently from other types of information. In other words, that DNA ancestry test results are inherently different from traditional genealogical records, and thus must be treated differently.
>
> Indeed, if DNA ancestry test results are *not* inherently different from traditional genealogical records, then the answer to the question of whether to "publish results that might affect relatives who have not released rights" would be the same for DNA as it is for census records, land records, tax records, and other genealogical records.
>
> If DNA ancestry test results *are* somehow different, then the editors are correct in their argument that we need standards for test results that we don't necessarily need for other types of records.

Bettinger's distinction is a valid one and he specifically limited it to "ancestry tests." The question raised in this discussion is also one for which there may be no one "correct" answer. Perception is all important and it is clearly in the eye of the beholder.

In the run-up to the passage of the Genetic Information Nondiscrimination Act (GINA) in 2008,

> [b]oth the U.S. Chamber of Commerce and America's Health Insurance Plans (AHIP) have countered that there is no convincing evidence that employers or insurers engage in genetic discrimination and that federal legislation to prohibit discrimination based on genetic information is unnecessary.[8]

In spite of these protestations, the bill prohibiting such discrimination passed in the Senate unanimously and in the House of Representatives by a vote of 414 to 1. It was then signed by President George W. Bush. Although it could be debated that there was "no convincing evidence" that this law was necessary, the U.S. Congress and President Bush apparently believed otherwise.

This action was taken at least in part because the Congressional Research Service had reported the following data:

- Sixty-eight percent of Americans are concerned about who would might access to their personal genetic information.
- Thirty-one percent state this concern would prevent them from having a genetic test.
- Sixty-eight percent agree that insurers would do everything possible to use genetic information to deny health coverage.[9]

It is important that both genealogists and others from whom we need to solicit test samples reach some consensus on whether the information in our genomes that we test for genealogical research is in any way exceptional and whether it requires special level of confidentiality.

DNA RESEARCH IS SOMEWHAT DIFFERENT

Genealogical research using DNA is fundamentally different from genealogical research using static documents as well as from DNA research for medical purposes in at least one regard. For family history research, the data collected from one's genome is genealogically meaningless in and of itself. Information from one's genome takes on genealogical meaning only when it can be compared with information from other individual genomes. That, in turn, can be accomplished only by convincing others to get tested and/or to share their test results. To gain this access requires building trust levels, including an expectation of confidentiality, similar to those required in the best professional/client relationships.

Library ethicist Jean Preer describes the importance of confidentiality extended by professionals:

> The protection of confidential information is one of the oldest ethical values and one of the most commonly shared among the professions. Essential to the relationship between doctor and patient, clergy and confessor, and lawyer and client, the duty to keep in confidence what is learned from or about a person in the course of providing service is a hallmark of professionalism. In becoming professionals, librarians somewhat belatedly incorporated the ethical obligation to protect confidences without fully realizing the extent of its reach or its implications for practice.[10]

Genetic genealogists today often unwittingly find themselves facing the same challenges set forth in Preer's last sentence.

DNA testing for genealogical research crosses into unknown territory almost daily—both genetically and ethically. We can pretend that most of these concerns relate solely to the medical application of genetics, but to do so is to ignore information that in many cases could be of great importance to individuals and their families. Genetic information has the potential to influence future behavior in ways more profoundly than information viewed only in more static documents—even the death certificates of close relatives. Many of these issues lead us beyond the scope of this book. A fuller discussion will have to await subsequent volumes. Those of you wishing to explore how some of these decisions can have very personal consequences may wish to read Jeff Wheelwright's 2012 book, *The Wandering Gene and the Indian Princess: Race, Religion, and DNA.*[11]

A NEW ETHICAL PARADIGM

What do a librarian, a hacker, and a person who takes a DNA test have in common? Michael Scherer in a recent *Time* article wrote of such a person, "He believes above all that information wants to be free, that privacy is sacred and that he has a responsibility to defend both ideas."[12] At first reading, those two concepts may seem in total opposition. However, when considering the moral and ethical issues surrounding the appropriate use of information derived from DNA testing, the situation can become much more complicated than the one described by Scherer.

Scherer was focusing only on hackers. I included the librarians in the previous section because I was a librarian for 35 years. In the last decade and a half of that career, I taught a class called "Ethics in the Information Age." It was a course for those students preparing for library careers or careers in web design in which their main goal would be to help others solve their information problems. In organizing the content for the class, I developed some models that I now realize are useful as guidelines in considering how we should treat the information disclosed by DNA tests. Discussion of these will make up much of the balance of this chapter:

- In ethical dilemmas, "rights" are often in collision.
- "Nothing is more dangerous than an idea when it is the only one we have" (French philosopher Alain).[13]
- Control and accountability are zero-sum games among the players in any situation.
- Reasonable people diligently attempting to be ethical can arrive at different and conflicting conclusions.

Preer says, "Ethical behavior does not mean adherence to strict rules of behavior but rather the application of values in situations where there is no fixed rule or clear right answer."[14] She could be describing the fluid situations in which we as genetic genealogists often find ourselves. Have we developed personal ethics that can effectively guide our behavior when we do not think anyone else is watching?

"RIGHTS" IN COLLISION

The conflict about ownership of information from within individuals can best be understood by examining various "rights" that are often asserted by one or more of the players who claim such rights. For simplicity of discussion, these rights/needs will be limited to the following four:

- The right to access information
- The right to keep information private
- The right to own and benefit from information
- The right to security by controlling information

In Figure 7.1, as one of these rights receives more emphasis, less importance can be given to other rights. For example, the impact of the terrorist attacks on September 11, 2001, resulted in a greater push for protection/security, indicated by the arrow at the bottom of the diagram. Greater emphasis on security significantly changed the balance in the United States away from individual rights, at least temporarily, by emphasizing the nation's perceived need for increased security. The revelations in 2013 by Edward Snowden of secret, widespread spying both at home and abroad by the National Security Agency (NSA),[15] remind us that this new balance point may need some readjustment. We are still as a nation trying to come to grips with this brave new world. As these modifications of policy take place, the box representing "the right thing to do" (Figure 7.1) shifts and forces one or more of the other arrows to shrink. It is a *zero-sum game*.

As shown in Figure 7.1, the "right thing to do" shifts as emphasis is moved from one "right" to another. Those who try to define their own behavior and the behavior

Figure 7.1
The "right thing to do" shifts as emphasis is moved from one "right" to another

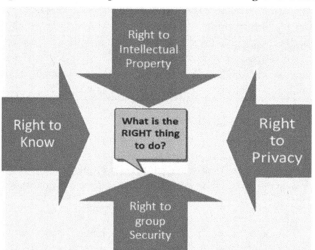

of others from the perspective of only one of these "rights" are doing all of us a disservice. In following such a path, they often discredit their own point of view by pushing their single-theme agendas well beyond the point where they can be considered to be credible. In ethics, no single principle, no matter how sacred, should be allowed to push all other concerns from our minds.

Right to Read Information from One's Genome

One of the strongest motivators in both my professional life and in my personal life is to learn new things. With graduate degrees in both history and library science, I have been trained to seek out information—dare I say "truth"? Sandwiched in between those degrees were my four years spent as a special investigator for the United States Air Force. During my four decades as a librarian and educator, my role was to find information and to help others learn to find it for themselves. But does the pursuit of truth trump all other values? Is there also a right not to know or to prevent others from knowing? Are there legal limits? Are there ethical limits? Are these limits different?

During the first decade of the 21st century, sharp differences of opinion arose regarding whether individuals should have the right to "read" the information coded within their cells. Many felt that the information could be dangerous if it was unmediated by a medical professional or genetic counselor. Corporations attempted to "patent" specific genes that have great medical significance so that they could monopolize and monetize tests on those genes. In 2011, I felt compelled to blog about the "right to own our own bodies"; my argument in this posting, with slight updates, is paraphrased here.

One of the most important human rights issues of the 21st century may be the right to have access to information contained in our own bodies. We are generally acknowledged to own any ideas we express that can be recorded in tangible form (copyright). However, we do not seem to have clear rights of ownership to more tangible products of our bodies. Body pieces that are removed during medical procedures, information gained during medical testing, and information contained in our DNA seem to fall into a category of information from which some paternalistically believe they must protect us.

After four decades as a librarian, I have the professional ethic that information is power and that my role as a professional is to help clients find the information they need to make informed decisions about all aspects of their lives. This is a somewhat different paradigm than the prevailing one in the medical arena between physicians and their patients.

The scientific and medical communities contain very diverse opinions.

Scientists and physicians want to use health information for their own research or diagnostic purposes, but some are reluctant to share it with the donors of that information. Such researchers are afraid that we might object to their use of these data, which they believe they can exploit for the benefit of society. Some physicians do not appreciate the value of their patients having information on medical issues, particularly when the physicians do not have the knowledge to "fix" the problems. Both of these positions are short-sighted and paternalistic.

Fortunately, such positions appear to be shifting toward the view that the physicians and patients can be productive partners in managing health-related life decisions if both are empowered with the best available information.

I had been developing my own opinion on these topics over the last few years as I pondered my own chances of developing Alzheimer's disease. My mother was afflicted by late-onset dementia that may have been a form of Alzheimer's. The prevailing view of the medical community seems to be that since doctors cannot successfully reverse or arrest such a disease, there is no benefit for me to find out whether I might have an elevated risk of developing it. In fact, they believe that such knowledge would unduly raise my anxiety level and lead to depression. A few recent studies suggest that any such anxiety and depression would be short lived and that those who do not test positive for increased risk will receive a positive attitude adjustment from the results.

Even if the medical community is not yet equipped to successfully treat me for such diseases, I can use such diagnostic knowledge to make informed decisions about my life and the environment in which I wish to live it. For example, if I believe myself to be at increased risk of Alzheimer's, I may wish to choose to live in an extended-care retirement community of my own choosing rather to inflict the burden of such a choice on my children with the potential guilt this may place on them.

With those concerns in mind, I have been tested for one gene marker that is thought to be associated with Alzheimer's. I fully understand that this is only one indicator and that by itself it does not control my fate. I was relieved to find that on the basis of this one indicator, my likelihood of developing Alzheimer's is only about 3 percent while that of the general population is about 7 percent. At the same time, my wife was tested along with me and discovered that she is at an increased risk of about 14 percent, even though she has no family history of Alzheimer's. Her result does not seal her fate. She still has, based on this one indicator, only a one in seven chance of developing this disease. However, based on this knowledge, we can make decisions about our lives that have little to do with the ability or inability of the medical profession to "fix" Alzheimer's disease.

Individual citizens need to have complete access to any medical test conducted on them and to be able to control information coming from within their bodies. For those of you who may have an interest in some of the ethical issues involved, I would recommend the thought-provoking book *The Immortal Life of Henrietta Lacks* by Rebecca Skloot.[16]

While that debate still goes on, the tide has clearly turned but the struggle is not over. Individual rights are asserting themselves. A year and a half after I blogged the "right to own our own bodies" manifesto, more than 1 million individuals had tested their DNA through direct-to-consumer tests and a unanimous U.S. Supreme Court reversed decades of legal precedents by ruling that it was no longer possible for anyone to claim ownership of all testing of a specific part of the human genome.

Supreme Court Gives Us Ownership of Our Genes

By now you have probably heard about yesterday's U.S. Supreme Court decision in the case of *AMP v. Myriad Genetics*. In that decision a unanimous court ruled that corporations could not patent a naturally occurring human gene even if they discovered its location in the human genome. ...

That was a huge victory for those of us who are optimistic about the promise of personalized genetic medicine in which our own particular genetic makeup is used to both diagnose and tailor treatment of our human disorders.

After the decision was announced, Myriad made a statement that other patents would enable the company to maintain its position. These would include its proprietary database that allows Myriad to interpret the results of its BRCA [breast cancer gene] testing. All of us who have taken genetic genealogy seriously have learned that DNA results, taken by themselves, have little meaning. It is only when these results can be compared with a large number of other results that meaningful interpretations can be made.

Other labs are already stepping forward to offer alternatives to Myriad's monopoly-supported pricing. What Myriad was charging about $4,000 for is now available at the drastically reduced price of $995 at a respected and accredited laboratory. ...

In a press release picked up by the *Wall Street Journal*, the Houston-based company Gene by Gene announced the availability of BRCA testing in the United States that it had previously only been able to offer abroad. Many of you are already customers of Gene-by-Gene through its Family Tree DNA (FTDNA) tests for family history information.

The developments of the last two days make BRCA testing much more affordable and probably are only the beginning of what competition in the marketplace will do to make genetic testing a routine part of our medical care.

Thank you to the U.S. Supreme Court for allowing us to own our own genes!

Did I go a little over the top in that post? Did I allow my euphoria over the freeing of my genes to cause me to lose my sense of perspective and to allow my ethical pendulum to swing too far off center? Perhaps. Such fluctuations happen to most of us from time to time. Is Gene-by-Gene's $995 product really the same as Myriad's $4,000 test? As I have pointed out elsewhere, the laboratory testing of specific locations on our genomes is not the real product that drives genetic genealogy. Rather, it is the ability to compare our results with those of others. The real value is in the databases and in the interpretation of our results. This is similar to medical tests such as those related to the BRCA gene. We need to evaluate more than just the retail price when making our choices.

As I write this chapter in 2014, we are witnessing the legal aftershocks that continue to rattle the genetic testing landscape. The rupture of the legal fault line initiated by the landmark *AMP v. Myriad Genetics* decision continues to drive what may be a lengthy period of seismic readjustments. Suits and countersuits by the principals dot the legal landscape. We are gradually edging toward a new equilibrium in the legal and ethical battle between the individual's right to know and the intellectual property rights of entrepreneurs. Myriad and Gene-by-Gene have reached an out-of-court settlement that will allow Gene-by-Gene to continue to market its BRCA test outside the United States but will restrict the company from selling that product inside the United States until early 2016, when the Myriad patent expires.

Right to Be Left Alone

By now, you will have surmised that I favor the right of individuals to be able to read the information encoded in their cells. At the same time, individuals should have the right not to read that information. If they do read the information themselves, they should have the right to restrict who else can read it.

When relatives or potential relatives are approached to give a DNA sample for documenting family history, a number of concerns are frequently voiced. These include some that are quite rational and some that are unfounded. Even those worries that on the surface sound irrational may mask unarticulated concerns that are valid. All can be barriers to research unless they can be resolved. Logic often has little to do with an individual's apprehensions. If that is the case, logic will not dissolve these apprehensions. Commonly voiced concerns include the following:

- "I don't want the government to find out what my DNA is. It's none of their business."
- "I'd rather not know about health problems in my future—particularly if I can't do anything about them."
- "If I should be discovered by some long-lost half-sibling, will there be inheritance issues?"
- "I don't want my insurance company to raise my health insurance rates."
- "I get nauseous when I see blood."
- "Why should I subject myself to something that is painful?"

Some concerns are simply based on fear of the unknown. Some of these fears can be overcome with a little basic information. However, we must be careful not to oversell the answers:

- Government employees generally have better things to do than to collect DNA information on law-abiding citizens. DNA taken for genealogical purposes cannot be admitted into evidence in court because of the "chain of custody" principle. No one verifies the identity of the person giving the sample. Therefore, in a legal sense it cannot be established that the tested sample came from a specific individual. Nevertheless, there has been at least one documented case where a forensic genealogist working with police on a cold case went fishing in a genealogical database after giving misleading information to the surname project administrator about the reason for her interest. Any information derived in that manner might help to develop an investigative lead but could never be accepted as evidence in a legal proceeding.
- Some people live in the moment and do not want to worry about the future. This is a personal choice that should be respected whether or not we agree with it. Of course, some health hazards can be addressed by physicians; at the same time, given the current state of medical practice, some cannot. If there currently is no cure, some people, including some medical practitioners, would rather not know about the health risk.

- President George W. Bush signed into law the Genetic Information Nondiscrimination Act (GINA) in 2008. The National Institutes of Health (NIH) website says in part,

> "Genetic discrimination occurs if people are treated unfairly because of differences in their DNA that increase their chances of getting a certain disease." ... [GINA] is a new federal law that protects Americans from being treated unfairly because of differences in their DNA that may affect their health. The new law prevents discrimination from health insurers and employers.[17]

> Note that "The law does not cover life insurance, disability insurance and long-term care insurance."[18] Some states, notably California, go further and add genetic discrimination to the list of categories (e.g., race, gender) against which discrimination is unlawful. In California, these protected areas include access to housing, education, business licenses, business establishments, and state-funded activities and programs.[19] The degree of protection from these laws varies widely among the different states. "The federal law sets a minimum standard of protection that must be met in all states. It does not weaken the protections provided by any state law."[20]

> However, it is possible there might be inheritance issues if a long lost half-sibling should emerge. The general principle in the United States seems to be that any such claims end if there is a legal adoption. However, this may not be the case with unknown illegitimacies that do not lead to a legal adoption. State laws vary. If this becomes a concern you should consult legal counsel. This issue did surface with one cousin who was weighing the pros and cons of testing to help resolve the parentage of newly emerged cousin Jim described in Chapter 5.

> Almost all direct-to-consumer genetic tests do not involve drawing blood and are no more invasive than rubbing the inside of one's cheeks with something like a cotton swab. Some rely on spit collected in a test tube. In my own experience, these tests have yielded the same genetic information as tests conducted in a health clinic involving the drawing of blood.

- Cheek swabbing is painful only if one rubs too hard (rare). In that case, a little blood might be drawn inadvertently. Spitting is boring, not painful.

It seems that most of the reservations of reluctant potential DNA contributors are without rational justification—at least from the perspective of eager genetic genealogists. At the same time, no matter how important we may perceive samples of their DNA to be for recreating an accurate family history, we must respect their right to privacy.

The Right to Privacy

Most of us can agree that certain things should remain private, at least when they relate to us. Governments and corporations also advocate for their own privacy to shield their activities from public scrutiny.

Technology ethicist Richard Spinello differentiates between many different kinds of privacy.[21] Among them are the right to be left alone and the right to control access to our own information. Most of us would agree that personally identifiable information such as Social Security numbers, the maiden names of mothers, and the dates of birth of living individuals can create havoc when they fall into the hands of the unscrupulous. Is there similar danger in publicly exposing information from inside our genomes?

Most of us would acknowledge that, at least in some situations, we behave differently when we believe others are watching. This phenomenon leads libertarian philosopher Joseph S. Fulda to argue, "If no one knows what I do, when I do it, and with whom I do it, no one can possibly interfere with it[; and] that a society cannot be free if citizens do not have a right to privacy. Privacy is essential because a government that is ignorant of an individual's thoughts and deeds cannot act to impinge on his or her rights."[22]

To some, like former U.S. Supreme Court Justice Louis Brandeis, privacy means "The right to be left alone—the most comprehensive of rights, and the right most valued by a free people."[23] The meaning of privacy has evolved since the days depicted in *Clan of the Cave Bear*, when everyone huddled up in the chamber of the cave. Daniel Boone was alleged to have felt that civilization was encroaching on his privacy when, standing in the clearing of his cabin, he could see the smoke from a neighbor's chimney. At that point, he knew it was time for him to move farther west and deeper into the wilderness. To others, privacy means the right to protect our persons and those of our loved ones. Closely related to this is the right to protect our physical property and our ideas.

Right to Be Forgotten

To this bundle of privacy rights, our European neighbors are in the process of establishing a new one—the right to be forgotten.

- Nine out of 10 Europeans (92 percent) say they are concerned about mobile apps collecting their data without their consent.
- Seven out of 10 Europeans are concerned about the potential use that companies may make of the information disclosed.[24]

To translate these concerns into public policy, in October 2013 the European Union (EU) Committee for Civil Liberties, Justice, and Home Affairs (LIBE) approved regulations designed to "put citizens back in control of their data."[25] The principal parts of these proposed rules are as follows:

- *A right to be forgotten.* When you no longer want your data to be processed and there are no legitimate grounds for retaining it, the data will be deleted. This is about empowering individuals, not about erasing past events or restricting freedom of the press.
- *Easier access to your own data.* A right to data portability will make it easier for you to transfer your personal data between service providers.
- *Putting you in control.* When your consent is required to process your data, you must be asked to give it explicitly. It cannot be assumed. Saying nothing is not the same thing as saying yes. Businesses and organizations will also need to inform you without undue delay about data breaches that could adversely affect you.
- *Data protection first, not an afterthought.* "Privacy by design" and "privacy by default" will also become essential principles in EU data protection rules. This means that data protection safeguards should be built into products and services from the earliest stage of development, and that privacy-friendly default settings should be the norm—for example, on social networks.[26]

In the immediate wake of the Snowden revelations of widespread NSA monitoring of communications,[27] this proposal was approved in committee by a vote of 49-1 and moved on for final adoption as a European-wide standard. If approved, it would establish a framework for European companies as well as for foreign companies operating in Europe. It remains to be seen how these principles might affect the operations of governments, corporations, and individuals. However, it may be some time before we have a chance to find out, because the top echelons of European leadership seem less eager than their underlings to adopt this new privacy standard.

Fourth Amendment

In the United States, the legal bases for our privacy rights flow from the Fourth Amendment, although you will not find the word "privacy" in it: *"The right of the people to be secure in their persons, houses, papers, and effects, against unreasonable searches and seizures, shall not be violated, and no Warrants shall issue, but upon probable cause, supported by Oath or affirmation, and particularly describing the place to be searched, and the persons or things to be seized."*[28]

ETHICS IN 3-D

Earlier in this chapter you were introduced to four competing rights. However, ethics is more complex than can easily be diagramed on a two-dimensional sheet of paper. Any ethical action has at least two "players." One may be active and the other passive, or both may be active. Each of them may see themselves as possessor of any or all of the four competing rights. Matters get even more complicated when we look at who these players may be.

The Players

Within the context of this chapter, I will oversimplify this complex process and assume that all the players in this drama can be lumped into three categories: individuals, corporations, or governments (Figure 7.2).

The three major groups shown in Figure 7.2 are players involved in a zero-sum game focusing on who controls the information in our genes. What each of these players does to impede or encourage the flow of information from our genes has an impact on the other two. Each of the three groups of players will assert that they have certain "rights." One rule of this game is that any one of these three groups can enhance its own rights but only at the expense of one or both of the other groups of players. The remaining players must, in turn, give up some of their rights.

Science fiction writer and privacy advocate David Brin observes that whenever a conflict arises between privacy and accountability, players demand the former for themselves and the latter for everybody else. He applies this rule of thumb to governments and corporations as well as to individuals. For example, governments want privacy for their own actions but insist on accountability for individuals and corporations. Individuals and corporations want privacy for themselves and accountability for everyone else. *"Each of us understands that knowledge can be power. We want to know as much*

Figure 7.2
Who controls our genes? Three major groups of players compete

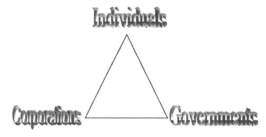

as possible about people or groups we see as threatening . . . and we want our opponents to know little about us. Each of us would prescribe armor for 'the good guys' and nakedness for our worst foes."[29]

Try to visualize the location of a piece of genetic information as being somewhere within the triangle depicted in Figure 7.2. The closer it is to one of the corners of the triangle, the more it is controlled by the player at that point of the triangle. If a player has total control of an item of information (items of information located completely in one of the three corners), that player alone can decide to keep it private, to share it, to sell it, or to give it away. Other players can try to gain control. However, it is a zero-sum game. As one player gains control, another loses privacy and ownership. Such a situation causes players to see their options in terms of a *win-lose* proposition. That is, one player can win something of value only if another loses something. Within an ethical context, one would hope all players would strive to achieve a fair balance between all the players so that a *win-win* scenario is approached, but theoretically that is not possible. At best, an equitable balance of rights is achieved.

Genetic information is power, and power is freedom. The freedom or power to act in perceived self-interest by any of these three groups is constrained or counterbalanced by the freedom or power possessed by the other two entities. Among the three entities there is a finite or fixed amount of freedom of action and a finite or fixed amount of control and accountability. For example, as the power of government increases, the freedom of corporations and/or of individuals decreases. All players seek to maximize freedom of action for themselves and to minimize their accountability to other players. The recent U.S. government crackdown on corporate accounting practices and executive compensation as the result of abuses at Enron, WorldCom, AIG, and other corporations is an example of how the government may increase its power by restricting the freedom of corporations. The USA Patriot Act is an example of how the government may increase its power by restricting some freedoms of individuals. Conversely, several recent court decisions have restricted the power of the government to regulate the Internet, thereby increasing the freedom of individuals and corporations. GINA constrains the ability of government and corporations to act on the basis of an individual's genetic information, thereby empowering individuals to inquire into their genetic makeup.

And then there is the U.S. Food and Drug Administration (FDA). Its role is still evolving as to what it should be doing to protecting consumers from the information within us and the interpretation of that information. Are these medical devices which should be regulated? We all would benefit if that role could be resolved quickly with the relatively small amount of information provided by 23andMe. Soon we will have to be dealing with whole genomes and the tiny amount of information tested by 23andMe will seem infinitesimally small.

The balance is constantly shifting. As one group gains, one or more of the others lose— hence the zero-sum game. It is sort of a three-way teeter-totter.

Where should the balance point of freedom be located within the triangle in Figure 7.2? Each of us may arrive at a different answer to this question, and on different topics. One of my goals is to make sure that you realize that each group depicted in Figure 7.2 can make very compelling arguments as to why its power needs to be increased. They usually rationalize that outcome as being in the interest of all of us. In reality, their power can be increased only if the other groups give up some of the freedom they now enjoy.

Figure 7.3
Selected Internet stakeholders (Reprinted with permission.)

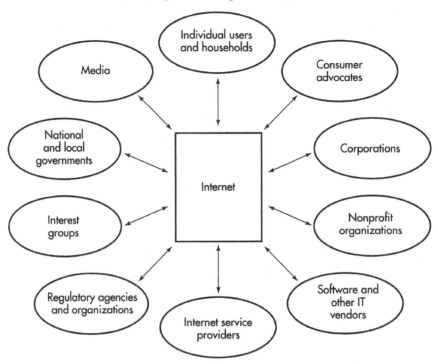

Other Stakeholders

As you probably realize, real life is much more complex than my simple three-cornered diagram. In his first edition of *CyberEthics*, Spinello included the graphic shown in Figure 7.3 to illustrate the variety of parties who try to influence behavior in cyberspace.[30]

You are encouraged to substitute the word "genome" for the world "Internet" in this diagram. I think Spinello dropped this illustration from more recent editions of his text because he believed it did not begin to convey the true complexity of the competing stakeholders. However, the chart is still instructive as a starting point for beginning to understand the current environment surrounding genetic genealogy. To do justice to this environment, many of Spinello's groups would have to be broken into subgroups—particularly the many and often competing interests within the medical and academic research communities.

COLLISION OF FREE INQUIRY AND PRIVACY

The freedom to inquire and the right to privacy often collide. Those inquiring often feel they *need* information that others may not wish to share. This is certainly true in family history research. Do these rights also support each other? The right of a free press to inquire cannot flourish unless the press is afforded some privacy to inquire. In library environments, patrons are assumed to be protected by anonymity as they inquire into any topic about which they are curious. This anonymity supports their free-dom to inquire.

Family historians wish to connect with other relatives. However, to secure the spit samples of some individuals, it may be necessary to promise at least minimal levels of anonymity. Many wish to test for health-related reasons but do not wish to share family connections. Others are happy to share but want to be protected from inquiring genealogists because they believe they know little about their family roots and have nothing to contribute to such dialog. In the latter case, they are happy to allow their DNA profiles to be used as long as trusted extended family members manage their accounts and answer those sticky genealogical questions.

Genetic genealogy is a powerful new tool, but also a double-edged sword. Not only can it bring us the joy of finding new cousins and even parents who may be delighted to be discovered, but it can also expose all members of those extended families to information about relationships and health conditions about which they might choose not to know. Once the gene is out of the test tube, it may be impossible to contain the flow of information from washing over even second or third cousins.

Even when personally identifiable information is supposedly removed from DNA databases, it may be possible for a motivated and skilled investigator to ferret out the identities of some donors.[31] We are left with a balancing act. For example, is the privacy of a sperm donor more important than the need of the resulting children to know their medical history? One had a choice in whether to become related; the others did not.

Americans give very mixed signals as to whether they really care about their online privacy. Like the Europeans mentioned earlier in this chapter, they tell pollsters that on-line privacy, at least in the abstract, is a big concern—but their actions contradict these findings. Intimate details are posted on social media. Only time will tell if government legislation such as GINA will prove to be a safe and effective tool to allow us to disclose our genetic information.

WHO OWNS YOUR DATA IN THE ERA OF "BIG DATA"?

The question of genomic privacy in the era of "big data" is becoming interlaced with a question about who owns our data. This link was made back in the pre-Snowden days of early 2013 by Gus Hunt, the chief technology officer of the Central Intelligence Agency (CIA). He told a tech conference:

> "The value of any piece of information is only known when you can connect it with something else that arrives at a future point in time," Hunt said. "Since you can't connect dots you don't have, it drives us into a mode of, we fundamentally try to collect everything and hang on to it forever."
>
> "Technology in this world is moving faster than government or law can keep up," he said. "It's moving faster I would argue than you can keep up: You should be asking the question of what are your rights and who owns your data."[32]

As genetic genealogists, it is in our interest for DNA testing companies to have robust databases with which we can compare our genomes. Generally, bigger is better in this field. These databases may be the most important service we are being offered. They also become important intellectual property for the companies.

What Is Intellectual Property?

Copyright expert Mary Minnow[33] defines intellectual property as follows: "Intellectual property is the intangible personal property that results from mental processes. There are four types of intellectual property that have legal protection in the U.S.: copyright, trademark, patents and trade secrets."[34] Copyrights and patents have their basis in the U.S. Constitution:

> "*Congress shall have the power to promote the progress of science and useful arts, by securing for limited times to authors and inventors the exclusive right to their respective writings and discoveries.*" (Article 1, Section 8, U.S. Constitution)

As an author, I appreciate the protection copyright gives to my writings. In the genetic genealogy arena, however, patents have been more at the forefront of efforts to protect intellectual property. Minnow describes patents as follows:

> Patents protect applied ideas, processes, and methods. Legal authority comes from the U.S. Constitution and Federal Law. The applied idea must be novel (no prior art), useful, and nonobvious. You must file a lengthy expensive application with the Patent Office (years and thousands of dollars with attorney fees). In order to get a patent, which grants you the exclusive right to your invention, you must disclose exactly how to reproduce it in your application, including detailed drawings if needed. Others may not use the invention even if they come up with the same idea independently. They must get a license from the inventor or wait until the patent expires. United States law grants utility patents for a term of twenty years from the date of application. Of note: in recent years, the U.S. Patent Office has been granting patents for "business methods" on the Internet.[35]

As a non-lawyer and non-geneticist, I find the world of genetic patents a daunting labyrinth of overlapping interests. However, we need to be aware that both copyrights and patents affect the ownership of our genetic data and our ability to read the information it contains.

The Emergence of "Big Data"

One of our recent buzzwords has been "big data." Without this phenomenon, which was described by Schonberger and Cukier in their best seller *Big Data: A Revolution That Will Transform How We Live, Work, and Think*, we would not be able to analyze the billions of bytes of DNA data available for hundreds of thousands of individuals. Following are snippets from these authors' thought-provoking book.

> [B]ig data is poised to shake up everything from businesses and the sciences to healthcare, government, education, economics, the humanities, and every other aspect of society.[36]
>
> Changes in the way we produce and interact with information lead to changes in the rules we use to govern ourselves, and in the values society needs to protect.[37]
>
> It leads to an ethical consideration of the role free will versus the dictatorship of data. Should free will trump big data, even if statistics argue otherwise? ... the age of big data will require new rules to safeguard the sanctity of the individual.[38]
>
> For decades an essential principle of privacy laws around the world has been to put individuals in control by letting them decide whether, how, and by whom their personal

information may be processed. In the Internet Age, this laudable idea has often morphed into a formulaic system of "notice and consent." In the era of big data, however, when much of data's value is in the secondary uses that may have been unimagined when the data was collected, such a mechanism to ensure privacy is no longer suitable.[39]

Although they build upon the values that were developed and enshrined for the world of small data, it's not simply a matter of refreshing old rules for new circumstances, but recognizing the need for new principles altogether.[40]

Schonberger and Cukier would have us rely on a system that punished those who abuse our data. But how can we hold governments and corporations accountable if we do not know which information is being collected, by whom, and how it is being used? Some have speculated that in the near future, we will all be identifiable in genetic databases, whether through our personal contribution or the contributions of our relatives. As this prediction comes closer to reality, how do we retain any semblance of a right to genetic privacy?

MEDICALLY RELEVANT INFORMATION

To tell or not to tell, that is the question—at least if medically relevant information is discovered as an unintended consequence of a genetic genealogy test. To date, the U.S. testing companies FTDNA and AncestryDNA have chosen not to go through the process of having their tests approved by the FDA and endorsed as "medical devices." 23andMe, in contrast, is seeking such a designation, although its path has not been smooth. These diverging business plans are the natural evolution of the origins of these companies.

Both AncestryDNA and FTDNA were formed to serve the genealogy community. In contrast, 23andMe's genealogy product emerged after the company was already involved in providing DNA testing direct to consumers. The original energy when Anne Wojcicki cofounded 23andMe was directed toward finding health indicators. This direction was accelerated and found a particular focus soon thereafter when Sergey Brin, cofounder of Google, discovered he carried a gene that dramatically elevates his probability of developing Parkinson's disease from about 1 percent to about 50 percent. A decade ago, Brin's aunt and mother had been diagnosed with the disease:

> At the time, scientific opinion held that Parkinson's was not hereditary.
>
> Even after the LRRK2 connection was made in 2004, Brin still didn't connect his mother's Parkinson's to his own health. Then, in 2006, his wife-to-be, Anne Wojcicki, started the personal genetics company 23andMe (Google is an investor). As an alpha tester, Brin had the chance to get an early look at his genome. He didn't find much of concern. But then Wojcicki suggested he look up a spot known as G2019S—the notch on the LRRK2 gene where an adenine nucleotide, the A in the ACTG code of DNA, sometimes substitutes for a guanine nucleotide, the G. And there it was: He had the mutation. His mother's 23andMe readout showed that she had it, too.[41]

Given this history, it is easy to understand why 23andMe's focus has remained on the relationship of genes to medical conditions. The company has offered 10,000 free testing kits to people who have received Parkinson's disease diagnoses so that it can create a large database through which to compare these individuals with other individuals who represent the general population (i.e., people who have not received such a diagnosis).

It is clear that 23andMe has a different mission and a different business plan than the companies that were formed for genealogical exploration as more narrowly defined. It is also understandable that many of those people who are in the 23andMe database are not eager to share their family histories—much to the chagrin of genetic genealogists.

To Tell or Not to Tell

No matter where you test or for which purpose, you may stumble upon medically relevant information if you probe deeply enough. Although our mitochondria have not yet been determined to carry much health-related information, they do give important indicators about both primary and secondary mitochondrial dysfunction. In the case of secondary dysfunction, the mutations on the mitochondria join with mutations in the nucleus to contribute to the disorder.[42]

One example of a primary dysfunction is the rare condition known as aminoglycoside-induced deafness. Some widely prescribed antibiotics can cause deafness—often permanent—in patients who carry the A1555G mutation. Research studies often exclude this gene from their panels because of patents claimed by Athena Diagnostics. This includes large-scale studies to correlate patient outcomes and drug reactions, such as Vanderbilt University Medical Center's PREDICT project, in which my wife and I are participants.[43]

Recent studies in China suggest that a C149T mutation in the mitochondrial 12S rRNA gene also plays a role in maternal inherited aminoglycoside-induced deafness as well as hypertension.[44] If you have the results of full mtDNA tests, you already have been tested for these mutations. The latter location was tested even if you have only HRV2 results.

If you as a genetic genealogist become aware that a family member or a member of your project has a medically significant mutation, should you tell the individual? Would it matter if no treatment is available given the current state of medical knowledge and practice? These are questions for which there are not universal answers that glibly can be applied in all circumstances. Nevertheless, awareness of such potential genetic minefields may help you avoid some regrettable missteps should you suddenly confront such situations.

FORENSIC GENEALOGY: PROTECTING A COMMUNITY IS A GOOD THING, ISN'T IT?

Most of us would be in favor of helping the police solve a homicide. By so doing, we contribute to making our community safer and enhance the peace of mind of its inhabitants. On a smaller scale, this is like being in favor of national security. But should we protect our communities by violating the confidentiality of many innocent individuals?

DNA results that are part of a genetic genealogy project are not considered to be legally protected. Were they part of an individual's medical records, they would be protected by the Health Insurance Portability and Accountability Act (HIPAA). "The HIPAA Privacy Rule provides federal protections for individually identifiable health information held by covered entities and their business associates."[45] If they were part of an educational record, "Generally, schools must have written permission from the parent or eligible student [older than age 18] in order to release any information from a student's education record."[46] Even what an individual checks out from a publicly funded library is legally protected in some states and cannot be revealed without a court-ordered subpoena. Which level of confidentiality should apply to DNA results?

A prominent forensic genealogist who often serves as a consultant to law enforcement approached a surname project coordinator. She had searched publicly available databases and found what appeared to be a possible match. She misled the coordinator into believing that she was pursuing a routine genealogy query and asked to be put in direct contact with participants in the project, but was rebuffed. She later provided information derived from the public sites in a press conference about the search to identify a killer in a three-decades-old cold case homicide investigation. Nothing illegal transpired. "Police will tell lies to uncover the truth; they will deceive, fabricate, and manipulate until the suspect surrenders—methods that are all perfectly legal."[47] But what about ethics? There was no indication that the killer was an imminent threat or was even still alive, yet all those who shared his suspected surname were, as a result of these actions, associated with his guilt, including a close friend of the murdered girl's family. No arrests have been made in the case and, as far as is known, the killer remains at large. The forensic genealogist refused to grasp why anyone would be upset that their privacy had been violated by her efforts to advance public safety. Conversely, she protected her own intellectual property. She wrote in a post in response to criticism, "I do not have to disclose the analysis I did. My technqiues [*sic*] are proprietary."[48]

A Single Ethical Tool?

Psychologist Abraham Maslow wrote, "I suppose it is tempting, if the only tool you have is a hammer, to treat everything as if it were a nail."[49] Would you take your car to a mechanic:

- Whose only tool is a hammer?
- Whose only tool is a screwdriver?
- Whose only tool is a pair of pliers?
- Whose only tool is a wrench?

Do you see a pattern in your answers? It is important to be able to choose from a variety of tools to solve automotive problems. The same is true in solving ethical problems. In ethical issues, as in much of life, we should beware of those who offer simplistic answers. French philosopher Alain warns, "Nothing is more dangerous than an idea when it is the only one we have."[50]

Even DNA Evidence Has Limits as a Tool

As seductive as the apparent certainty that comes from DNA matches in a *CSI* television show setting, there are limits to the conclusions we should draw. We are just beginning to compile examples of how best to curtail excesses in forensic DNA. Two examples illustrate cases where the science was not wrong, but the interpretation was wrong because of sample contamination. In the first example, German police wasted years and hundreds of thousands of euros pursuing a nonexistent serial killer because the DNA thought to have been retrieved from multiple crime scenes was, in fact, that of a woman who packaged the testing kits.[51] In a more recent example, a man was jailed for five months awaiting trial for a home invasion robbery and murder apparently because he was transported to the hospital by emergency medical technicians who later that evening responded to the scene of the crime and inadvertently transferred his DNA to that location.[52] Other leads that might have led to the apprehension of the real perpetrators in both these cases were ignored because of the assumed certainty of DNA evidence.

EMERGING ETHICAL ISSUES

"Big Data" and Genomic Information

Schonberger and Cukier wrote about how Big Data may come to genomic information:

> Data was no longer regarded as static or stale, whose usefulness was finished once the purpose for which it was collected was achieved. . . . Rather, data became a raw material of business, a vital economic input, used to create a new form of economic value.[53]
>
> [Big data leads to things] one can do at a large scale that cannot be done at a smaller one, to extract new insights or create new forms of value, in ways that change markets, organizations, and the relationship between citizens and governments and more.[54]

An analysis done in Finland of genomic testing companies points out, "the genome data collected from the customer volume can be utilised through the resale of anonymous cohorts. The genetic data is the company's critical key resource with regard to both producing the service and generating a revenue stream through the resale of the data."[55]

Retired scientist, entrepreneur, and genetic genealogist Walter Freeman offers the cynical view that the customer service profiles of current testing companies indicate that their real business models are to collect and analyze our aggregate genetic data so that it can be remarketed at great profit:

> The public is not their customer. Genetic genealogists certainly are not their customer, nor have they ever been. (Genetic genealogy is just a means to an end to encourage a few more customers to sign up, hence the arm's length, low customer contact, relationship, though they try to make it look like a party full of like-minded people.)[56]

It remains to be seen whether any of these companies will turn to data mining of the anonymized DNA records of our individual test results to generate an additional revenue stream in the big data era. Although we may not have read the fine print on the consent forms we signed, this is not a use most of us envisioned for our samples.

Informed Consent

Generally, we have tried to test individuals only after they have given informed consent. That concept has grown out of a patient's agreement to surgery.[57] Writing about genomic research in 2010, Amy L. McGuire and Laura M. Beskow expanded this definition:

> Informed consent is a cornerstone of the ethical conduct of research involving humans. Based on the ethical principle of respect for persons, the goal of informed consent is to ensure that subjects are aware of the risks and potential benefits and make a voluntary

decision about participating in the research. However, advances in genetic and genomic research—in particular, the increasing emergence of large-scale population studies and genomic databases—have challenged traditional conceptions of informed consent.[58]

At the time of this writing, more than 1 million individuals had likely taken direct-to-consumer DNA tests. It is assumed that they did so after providing informed consent. However, these individuals may be a minority of all people who have their DNA or at least their DNA test results currently stored in databases. How many more of us have been tested? Thousands of persons accused of serious crimes have been tested and their results entered into "criminal databases." Some other kinds of DNA databases are less well known—namely, those holding the results of medical procedures. Almost all infants in the United States are routinely tested to screen for several conditions—some of them life threatening if not treated. In some states, these samples are destroyed after testing; in others, they are not.[59] Should parents (read "fathers") have access to the results when it is possible they are not the biological parents? Also, most individuals who have undergone surgical procedures or died in a hospital have had tissue samples removed. These samples seem to be stored indefinitely by preserving the tissue in paraffin blocks.

The ethics and procedures for using these kinds of DNA samples to determine family relationships were explored by Lennard Davis in his autobiographical *Go Ask Your Father: One Man's Obsession to Find His Origins through DNA Testing*.[60] Who should have access to the results? During his quest, Davis was not sure whether he was the biological or legal next of kin to the man who had raised him. Should these samples be destroyed after their medical use is complete? More questions and few clear-cut answers arise when we ponder this issue.

SITUATIONAL ETHICS

Each person considering ethical issues must decide whether there are absolute ethical principles that can be applied universally, or whether *situational ethics* might cause the correct answer to vary from one setting to another. Are there some absolutes upon which we all can agree? Is it possible that two people, each desperately trying to do the ethical thing, will adopt very different courses of action? In applied ethics, often it is necessary to decide what "right" is and what "wrong" is. However, the real tension arise in situations where two or more "right" concepts appear to be in conflict. We still must resolve the competing rights to access information, to keep information private, to own and benefit from information, and to be secure by controlling information.

SYNCHRONIZING OUR HEARTS AND OUR HEADS

With ever-increasing speed, we are being forced to decide on the appropriate uses for the flow of information from our genes. Even back in the relatively tranquil times of the 1970s, some were already feeling overwhelmed by having to deal with issues related to the appropriate use of technology. A character in the novel *Wild Card* wanted to stop the world until he figured it out: "Our first priority should be learning to live with the technology we have already, not acquiring more. Because, like the Sorcerer's Apprentice, we just ain't going to be able to handle it."[61] For today's generations, stopping the flood of information from our genes is not an option.

Since that sentiment was expressed almost four decades ago, the pace of technological change has accelerated at an ever-increasing speed. It is difficult for most of us to understand how fast this growth in technology is occurring, let alone to intelligently and ethically manage its uses for the benefit of humanity.

Increasing the flow of the information from our genes is the easy part. Deciding the purposes to which that flow should be applied is a problem that continues to perplex us. A fictional character pondered this question even before the World Wide Web exploded in the midst of our society: "When Garver allowed himself to dwell on it very long, he almost despaired. It was marvelous what man had learned to do with nothing more than an electric spark. But somehow, he felt as though man was also only the alchemist's apprentice."[62] Garver reflected further: "He knew a bit of God's technology, but he understood considerably less of the divine moral sense that would enable him always to use it wisely. As history had proved all too consistently over the millennia, man's head was still ahead of his heart."[63] And no relief seems to be in sight for any of us.

SIX WAYS OF KNOWING TRUTH

The faster things change, the more we need to keep in touch with the timeless principles that have guided humans to truth through the ages. An old philosophy professor of mine suggested the following six ways that we can arrive at truth. How many different ones do you use?

1. *Rational thought.* Truth comes from logically correct thinking. (Everything grows out of Rene Descartes's observation, "I think, therefore I am.")
2. *Common sense.* Truth comes from what we all know in common.
3. *Sensory observation.* Truth comes from what we can see, touch, smell, taste and hear.
4. *Intuition.* Truth comes from within the individual—a sort of sixth sense, inner voice, or consciousness.
5. *Scientific method.* Truth comes from rigorously testing hypotheses and sharing results with others to see if they can be replicated.
6. *Authoritarian faith.* Truth comes from an authority in which we have faith (e.g., the Pope, the Bible, Rush Limbaugh, Nelson Mandela, Dr. Laura, Dr. Martin Luther King, Jr., our parents, Osama bin Laden, the government, a professional body, a mentor, Confucius).[64]

Only if we use authoritarian faith as our prism for discerning truth are we guaranteed that the *legal answer* will be the same as the *ethical one*. This will be true, however, only if the authority in which we place our faith is the wisdom of the government. If we use any of the other five means of determining truth or use authority figures other than the government for guidance, we may at times find ourselves in situations where ethical and legal answers can be in conflict.

SUMMARY

We are still attempting to balance human values and technological possibilities. We are still trying to synchronize our hearts with what our heads can create. We will continue this struggle for the rest of our lives. Dante reminds us, "The darkest places in hell are reserved for those who maintain their neutrality in times of moral crisis."[65]

NOTES

1. Quotewave.com, http://www.quoteswave.com/text-quotes/392136, accessed September 27, 2013.

2. Viktor Mayer Schonberger and Kenneth Cukier, *Big Data: A Revolution That Will Transform How We Live, Work, and Think* (New York, NY: Houghton Mifflin Harcourt, 2013), 171.

3. Schonberger and Cukier, *Big Data.*

4. Schonberger and Cukier, *Big Data,* 6.

5. Judy G. Russell, "The Ethical Genealogist," presentation at Jamboree 2013, Southern California Genealogical Society, June 7–9, 2013.

6. Board for Certification of Genealogists (BCG), "Code of Ethics and Conduct," http://www.bcgcertification.org/aboutbcg/code.html, accessed July 25, 2013. The National Genealogical Society has published the following pages on genealogical standards. Every genealogist, whether beginner or expert, hobbyist or professional, should read them (accessed September 27, 2013):

- Standards for Sound Genealogical Research
- Guidelines for Using Records, Repositories, and Libraries
- Standards for Use of Technology in Genealogical Research
- Standards for Sharing Information with Others
- Guidelines for Publishing Web Pages on the Internet
- Guidelines for Genealogical Self-Improvement and Growth

7. Melinde Lutz Byrne and Thomas W. Jones, "DNA Standards," *National Genealogical Society Quarterly* 101 (December 2013): 243.

8. Nancy Lee Jones and Amanda K. Sarata, "Genetic Information: Legal Issues Relating to Discrimination and Privacy," *Congressional Research Service (CRS) Report for Congress* (March 10, 2008), http://assets.opencrs.com/rpts/RL30006_20080310.pdf, accessed March 18, 2014.

9. Jones and Sarata, "Genetic Information."

10. Jean Preer, *Library Ethics* (Westport, CT: Libraries Unlimited, 2008), 183.

11. Jeff Wheelwright, *The Wandering Gene and the Indian Princess: Race, Religion, and DNA* (New York, NY: Norton, 2012).

12. Michael Scherer, "The Geeks Who Leak," *Time* (June 24, 2013): 24.

13. Emile-Auguste Chartier, alias Alain.

14. Preer, *Library Ethics,* xiv.

15. Lorenzo Franceschi-Bicchierai, "Edward Snowden Is Runner-up for *Time*'s Person of the Year," *Mashable* (December 11, 2013), http://mashable.com/2013/12/11/edward-snowden-wins-times-person-of-the-year/, accessed December 31, 2013.

16. David R. Dowell, "Right to Own Our Own Bodies," *Dr. D Digs up Ancestors,* http://blog.ddowell.com/2011/11/right-to-own-our-own-bodies.html, accessed September 25, 2013.

17. National Institutes of Health, "Genetic Information Nondiscrimination Act of 2008," http://www.genome.gov/10002328#al-3, accessed September 6, 2013.

18. National Institutes of Health, "Genetic Information Nondiscrimination Act."

19. Jennifer K. Wagner, "A New Law to Raise GINA's Floor in California," *Genomics Law Report: News and Analysis from the Intersection of Genomics, Personalized Medicine and the Law,* http://www.genomicslawreport.com/index.php/2011/12/07/a-new-law-to-raise-ginas-floor-in-california/#more-6345, accessed September 6, 2013.

20. National Institutes of Health, "Genetic Information Nondiscrimination Act."

21. Richard A. Spinello, *CyberEthics: Morality and Law in Cyberspace*, 4th ed. (Sudbury, MA: Jones and Bartlett, 2011), 150–153.

22. Joseph S. Fulda, "A Loss of Privacy Harms Society," in *Opposing Viewpoints: Civil Liberties,* ed. Tamara L. Roleff (San Diego, CA: Greenhaven Press, 1999), 77–80.

23. Justice Louis Brandeis, *Olmstead v. U.S.* (1928).

24. Committee for Civil Liberties, Justice, and Home Affairs (LIBE), "LIBE Committee Vote Backs New EU Data Protection Rules," *Press Releases Database,* http://europa.eu/rapid/press -release_MEMO-13-923_en.htm?locale=en, accessed October 30, 2013.

25. LIBE, "LIBE Committee Vote."

26. LIBE, "LIBE Committee Vote."

27. Franceschi-Bicchierai, "Edward Snowden."

28. U.S. Constitution, Amendment IV, The Bill of Rights (1791).

29. Brin, David, "A Parable about Openness ... Followed by Some Thoughts on Privacy, Security and Surveillance in the Information Age," http://www.davidbrin.com/akademos.html, accessed September 25, 2013.

30. Richard Spinello, *CyberEthics: Morality and Law in Cyberspace* (Sudbury, MA: Jones and Bartlett, 2000), 33.

31. Gina Kolata, "Poking Holes in Genetic Privacy," *The New York Times* (June 16, 2013), http://www.nytimes.com/2013/06/18/science/poking-holes-in-the-privacy-of-dna.html?pagewanted =all&_r=0&pagewanted=print, viewed, accessed January 2, 2014.

32. Matt Sledge, "CIA's Gus Hunt on Big Data: We 'Try to Collect Everything and Hang on to It Forever,' " posted March 30, 2013, updated March 21, 2013, http://www.huffingtonpost.com /2013/03/20/cia-gus-hunt-big-data_n_2917842.html, accessed November 2, 2013.

33. In September 2010, Mary Minnow's nomination by President Barack Obama to serve a four-year term on the National Museum and Library Services Board was approved by the U.S. Senate. In that capacity, she will be able to advise leaders of the federal government in policy matters that involve libraries and museums—including copyright matters.

34. Mary Minnow, Info People Workshop handout, April 2002.

35. Minnow, Info People Workshop handout.

36. Schonberger and Cukier, *Big Data,* 11.

37. Schonberger and Cukier, *Big Data,* 171.

38. Schonberger and Cukier, *Big Data,* 17.

39. Schonberger and Cukier, *Big Data,* 173.

40. Schonberger and Cukier, *Big Data,* 17.

41. Thomas Goetz, "Sergey's Search for a Parkinson's Cure," *Wired* (July 2010), http:// www.wired.com/magazine/2010/06/ff_sergeys_search/, accessed September 10, 2013.

42. Patrick F. Chinnery, "Mitochondrial Disorders Overview," *Gene Reviews,* http:// www.ncbi.nlm.nih.gov/books/NBK1224/, accessed January 3, 2014.

43. David R. Dowell, "Genomes, Hype, and a Realistic Pathway to Personalized Medicine," *Dr. D Digs up Ancestors,* September 29, 2012, http://blog.ddowell.com/2012/09/genomes-hype -and-realistic-pathway-to.html, accessed September 25, 2013.

44. Hong Chen, "The 12S rRNA A1555G Mutation in the Mitochondrial Haplogroup D5a Is Responsible for Maternally Inherited Hypertension and Hearing Loss in Two Chinese Pedigrees," *European Journal of Human Genetics* 20, no. 6 (2012): 607–612, http://www.ncbi.nlm.nih.gov/ pmc/articles/PMC3355256/, accessed September 24, 2013.

45. U.S. Department of Health and Human Services, "Understanding Health Information Privacy," http://www.hhs.gov/ocr/privacy/hipaa/understanding/, accessed November 2, 2013.

46. U.S. Department of Education, "Family Educational Rights and Privacy Act (FERPA)," http://www.ed.gov/policy/gen/guid/fpco/ferpa/index.html, accessed November 2, 2013.

47. http://www.vanderbilt.edu/jotl/manage/wp-content/uploads/Khasin-cr_final_final.pdf, accessed September 28, 2013.

48. Colleen Fitzpatrick, "Re: Interview Tonight—KCPQ Channel 13 Seattle," post to ISOGG@yahoogroups.com, January 9, 2012.

49. Abraham H. Maslow, *The Psychology of Science* New York: Harper & Row, 1966), 15.

50. Emile-Auguste Chartier, alias Alain.

51. Claudia Himmelreich, "Germany's Phantom Serial Killer: A DNA Blunder," http://www.time.com/time/world/article/0,8599,1888126,00.html#ixzz2a9Y8MYsl, accessed July 26, 2013.

52. Osagie K. Obasogie, "High-Tech, High-Risk Forensics," *New York Times* (July 24, 2013), http://www.nytimes.com/2013/07/25/opinion/high-tech-high-risk-forensics.html?smid=fb-share, accessed July 26, 2013.

53. Schonberger and Cukier, *Big Data,* 5.

54. Schonberger and Cukier, *Big Data,* 6.

55. Antero Vanhala and Karita Reijonsaari, "23andMe: Genetic Tests as Entertainment and as a Data Gathering," *Direct-to-Consumer Genome Data Services and Their Business Models* (September 19, 2013): 14, http://www.sitra.fi, accessed November 5, 2013.

56. Walter J. Freeman, "Direct-to-Consumer Genome Data Services and Their Business Models," post to ISOGG@yahoogroups.com , September 24, 2013.

57. *Black's Law Dictionary Free 2nd Ed. and The Law Dictionary*, http://thelawdictionary .org/informed-consent/#ixzz2gNABSZ6F, accessed September 30, 2013.

58. Amy L. McGuire and Laura M. Beskow, "Informed Consent in Genomics and Genetic Research,"*Annual Review of Genomics and Human Genetics* 11 (September 22, 2010): 361–381, 361, http://www.ncbi.nlm.nih.gov/pmc/articles/PMC3216676/, accessed September 28, 2013.

59. American Civil Liberties Union, "Newborn DNA Banking," https://www.aclu.org/free-speech-technology-and-liberty-womens-rights/newborn-dna-banking, accessed September 30, 2013.

60. Lennard J. Davis, *Go Ask Your Father: One Man's Obsession to Find His Origins through DNA Testing* (New York, NY: Bantam Books, 2009).

61. Raymond Hawkey and Roger Bingham, *Wild Card*(New York, NY: Stein and Day, 1974), xiii, 116. In *Oxford English Dictionary* (OED).

62. David Lindsey, *An Absence of Light* (New York, NY: Bantam, 1995), 304.

63. Lindsey, *An Absence of Light,* 304.

64. A context for these ways of knowing truth can be found in Denise K. Fourie and David R. Dowell, *Libraries in the Information Age: An Introduction and Career Exploration,* 2nd ed. (Santa Barbara, CA: Libraries Unlimited, 2009), 195–196.

65. Dante Alighieri, *The Divine Comedy*, quoted in Dan Brown, *Inferno: A Novel* (New York, NY: Doubleday, 2013), 463.

8

How Do You Continue to Document Your Family Story? Continuous Learning

Have you already made the DNA connection? If so, do you have a plan for how you will continue to learn as much as possible from the information contained in your cells as well as the cells of your close and distant relatives? Do you have realistic goals for what you can learn?

In beginning or continuing your journey into genetic genealogy, remember that each of the four major testing programs has something unique to offer:

1. Geno 2.0 focuses on deep ancestry and on collecting and preserving disparate and unique DNA SNPs from indigenous peoples around the world.
2. 23andMe focuses on health data as well as genetic genealogy information from atDNA.
3. AncestryDNA offers atDNA testing and automatic searching for matches who have in-depth pedigree charts also posted there.
4. FTDNA offers atDNA testing and is the sole choice for those who wish to compare their yDNA or mtDNA in a robust database.

Even if you have tested through one or more of these services and mastered all the techniques of genetic genealogy introduced in this book, many paths of discovery await you in your journey to apply the growing set of tools of genetic genealogy and unlock and analyze additional information in your genome.

Where do you want to go next in this incredible 21st century of inner discovery? How do you get there? So much has already occurred in genetic genealogy since 2000—yet we still have so much to learn. To organize these tasks within an action plan, you need to take stock of which information you already have. Should your next step be to collect more data or to learn more techniques for interpreting it? This book has offered assistance in only the most basic kinds of DNA analysis for family history. What should be your next step? What should be your overall plan?

COLLECT AS MUCH DNA FROM MEMBERS OF YOUR FAMILY AS YOU CAN AFFORD

The high water mark of applying the rigors of the historical method to genealogy research in the last half of the 20th century and beginning of the 21st century has been the *Genealogical Proof Standard* (GPS) of the Association of Professional Genealogists (APG). Lecturer, writer, and editor Thomas Jones is one of the best-known 21st-century advocates of this kind of careful and thorough genealogy. He guides us through the use of the information we observe in our genomes:

> DNA testing gives family historians access to biological data via DNA records and reports. We must interpret these documents in the same way we interpret other kinds of complex sources (land records, for example). DNA samples that do—or do not—match are genealogically significant, but without documentary data DNA reports cannot help support or disprove any conclusion of relationship or nonrelationship. The GPS is as important in contexts using DNA results as it is in contexts without them.[1]
>
> Records—whether of DNA markers, instrument settings, oral statements, actions, or events—are much more useful because we can use them to verify our readings and interpretations. Examining DNA and hearing oral history are nearly useless activities, for example, if no one makes a record of what they saw or heard.[2]
>
> *Source* refers to an entire item, not the information or evidence within it. In other words, sources are containers, not contents. Sources include specific books, censuses, certificates, compiled and narrative genealogies, court packets, deed and will books, DNA records or reports, family Bibles, manuscripts and published volumes of vital records, newspapers, religious records, websites, and many other containers of genealogically useful information.[3]
>
> We also might obtain reports of various kinds of DNA tests. Exactly what we examine depends on our research questions, the hypotheses we need to test, what is available for the time and place, our knowledge and expertise.[4]

How much DNA should you test? Jones likens DNA to oral history: "Consistently make your first genealogical priority to do today what future generation cannot do tomorrow. . . . Gather Y-chromosome, mitochondrial and autosomal DNA samples for as many lines and markers as you can afford."[5] Every time members of the older generation of our families die without having their DNA recorded, we lose part of the history of our families, just as surely as we do when we fail to record the oral histories of our elders.

This does not mean that every family member needs to be tested with every kind of DNA test that comes on the market. Unless we have reason to believe there may have been non-paternity events or other unresolved mysteries, it may not be necessary to test the yDNA and mtDNA of multiple close family members. However, because of the less predictable inheritance patterns of atDNA and xDNA, we can almost always learn something even from testing multiple siblings and close cousins.

ADVANCES IN GENETICS

Most of the information in our genomes has yet to be fully explored. Among the four areas in our cells where our DNA is stored, our mitochondria host by far the smallest amount. As a result, for the last several years we have been able to read all the letters of this information. We are still a long way from being able to claim we understand all the information the mitochondria contain. However, with only 16,569 locations, mtDNA exists on such a size and scope that we can visualize the day when we will have decoded most of what this small portion of the inner us has to say. Because its inheritance pattern does not include recombination, this information is particularly useful to genetic genealogists. Also, because of the bounteous multiple copies in every cell, mtDNA is more likely to retain and share its information even in very old human remains.

We are rapidly expanding our exploration of other areas of our genomes that may unlock new chapters of our family histories. The nuclei of each of our cells contain much vaster warehouses of information carriers. For men, our Y chromosomes are our next smallest vessels of DNA. They contain approximately 3,500 times as many base pairs as do our mitochondria. It will be a long time before we know how many of these locations hold genealogically relevant information, but the magnitude of this challenge is clearly daunting. Big chunks of this chromosome are now being explored by both the Big Y test from FTDNA and the so-called Full Y test offered by multinational start-up Full Genomes Corporation. At present, these tests are not available to those without some disposable income, and they definitely are not recommended for novices. However, on a targeted basis and within the context of haplogroup and surname projects of discovery, these are fast emerging as exciting areas. The current Big Y and Full Y tests are discovering huge numbers of new SNPs that are beginning toallow us to fill in gaps in the deep histories of the migrations of our paternal lines. Some of these test results show promise in terms of helping us sort out our family ties even in the genealogical time of the last few centuries or the last few generations. It is possible to envision a time in the near future when many of us will be able to identify "terminal" ySNPs that will identify our paternal lines as specifically as did the coats of arms in days of old.

On the other side of the Atlantic, Chromo2 has been launched. As I have not yet received my personal results from this test, I suggest that you read Debbie Kennett's comprehensive two-part blog post on this option.[6]

You will want to monitor these developments to learn how they may apply to your own efforts to extend your family history.

THE $1,000 GENOME

For several years, one of the continuing promises repeated in media coverage of DNA testing is that the $1,000 test for our complete genome is just around the corner.[7] That price is realistically thought to be a psychologically important trigger point for genetic genealogists. Full genome sequencing can now be purchased for less than $10,000. Technological breakthroughs in both DNA sequencing and in the computing power needed to store and analyze results have led the way to the lower prices. In both of these areas, we have benefited enormously from the investments being pumped into those organizations investigating the promise of personalized genomic medicine. Both government agencies and private-sector investors in the massive U.S. health industry, more so than genetic genealogists, have been the driving force behind the rapid innovations related to unlocking our genomes.

PRESERVE IT

As Thomas Jones has pointed out, we need to think about the records of DNA tests as being different from the DNA itself. Both the DNA itself and records of the tests of DNA need to be collected, analyzed, and *preserved* so that they will be available to your children and grandchildren as they research their family histories.

Almost all DNA testing for family history purposes has been completed in the last decade; perhaps half of it has been done in just the last couple of years. Already many of those who have been tested are deceased. It is wonderful that we were able to collect their DNA while they were still with us, but sticky legal and ethical issues remain to be resolved. Who owns their physical DNA samples now that they are gone? Who has a legal right to access the records of any tests conducted on them? Few of us think about including our DNA among the items we wish to bequeath when we draw up our wills. We do not think about our DNA testing accounts when we provide our next of kin with lists and passwords to our bank accounts, investment accounts, and other such assets. Our physical DNA and the records of its analysis have the potential to become some of the most cherished intellectual property we can pass on to our posterity, but we must act to ensure they will be preserved and will continue to be accessible.

DEEPER ANALYSIS

Those of you wishing to develop skills in more complex methods of analyzing the information you already have will want to consider whether learning how to apply such tools as triangulation would be useful. Although a few simple applications of triangulation were hinted at in the examples given in previous chapters, a more complete exploration of the possibilities of this rather simple but powerful tool could be the subject of another book. The basic concept is that an unknown can be calculated if two sides of a triangle are known. One simple example in genetic genealogy is that if two (or more) lines from a common ancestor can be traced backward from the present, we can then discover more about that common ancestor than we could by knowing only one line. In such a case the whole can become much more useful than the sum of its parts. The application of this concept is limited only by your imagination *and* data.

Chromosome mapping is generally much more rigorous and requires a tremendous investment of time and a generous testing budget.[8] Multiple samples from closely related family members of at least two or three generations are required. However, with a little luck and a lot of perseverance, it is often possible to follow specific segments of atDNA as it recombines through time. This technique has potential for tracing the person from whom a particular health threat was inherited or a particular ethnic contribution was passed down.

CONTINUE LEARNING

This book has hardly scratched the surface of the possibilities of applying genetic genealogy for the purpose of extending your family history. If you wish to continue to learn, read additional books, articles, blogs, and listservs. A small selection of these resources is listed in the appendix. You should also take advantage of conferences, programs, and webinars sponsored by genealogical societies and testing companies. Check their websites and Facebook pages and other social media for opportunities.

You are joining a community in which no one can possibly know everything. We all learn from each other. I hope this book has raised as many questions as it has answered for you and that you are motivated to continue your journey toward becoming a more skilled genetic genealogist. There will always be new things to learn. Bon voyage!

NOTES

1. Thomas W. Jones, *Mastering Genealogical Proof,* National Genealogical Society Special Topics Series: #107, 2013, Kindle locations 265–268.

2. Jones, *Mastering Genealogical Proof,* Kindle location 351.

3. Jones, *Mastering Genealogical Proof,* Kindle location 354.

4. Jones, *Mastering Genealogical Proof,* Kindle location 689.

5. Thomas W. Jones, "Will Your Family History Have Lasting Value?", handout for presentation at RootsTech 2014, February 2014.

6. Debbie Kennett, "A First Look at the BritainsDNA Chromo 2 Y-DNA and mtDNA tests," *Cruwys News* (December 20, 2014), http://cruwys.blogspot.com/2013/12/a-first-look-at -britainsdna-chromo-2-y.html and "A First Look at the Chromo 2 All My Ancestry Test from BritainsDNA," *Cruwys News* (December 20, 2014, http://cruwys.blogspot.co.uk/2013/12/a-first -look-at-chromo-2-all-my.html, both accessed April 5, 2014.

7. Erika Check Hayden, "Technology: The $1,000 Genome," *Nature* (March 19, 2014), http:// www.nature.com/news/techno logy-the-1-000-genome-1.14901, accessed April 4, 2014.

8. Tim Janzen, Emily Aulicino, and Lisa McCullough, "Basics of Chromosome Mapping," presentation to GFO DNA Interest Group Meeting, July 27, 2013, http://www.gfo.org/intgrp/ chromo-mapping.pdf, accessed April 4, 2014.

Further Reading

BOOKS

Abraham, Carolyn. *The Juggler's Children: A Journey into Family, Legend and Genes that Bind Us.* Toronto, ON: Random House Canada, 2013.

Aulicio, Emily D. *Genetic Genealogy: The Basics and Beyond.* Bloomington, IN: AnchorHouse, 2014.

Brown, Dan. *Inferno: A Novel.* New York, NY: Doubleday, 2013.

Collins, Francis S. *The Language of Life: DNA and the Revolution in Personalized Medicine.* New York, NY: HarperCollins, 2010.

Davis, Lennard J. *Go Ask Your Father: One Man's Obsession to Find His Origins through DNA Testing.* New York, NY: Bantam Books, 2009.

Dowell, David R. *Crash Course in Genealogy.* Santa Barbara, CA: Libraries Unlimited, 2011.

Fourie, Denise K., and David R. Dowell. "Ethics in the Information Age." Chapter 8 in *Libraries in the Information Age: An Introduction and Career Exploration,* 2nd ed. Santa Barbara, CA: Libraries Unlimited, 2009.

Hill, Richard. *Finding Family: My Search for Roots and the Secrets in My DNA.* Richard Hill, 2012.

Jackson, Buzzy. *Shaking the Family Tree: Blue Bloods, Black Sheep, and Other Obsessions of an Accidental Genealogist.* New York, NY: Touchstone, 2010.

Kean, Sam. *The Violinist's Thumb: And Other Lost Tales of Love, War, and Genius, as Written by Our Genetic Code.* New York, NY: Little, Brown, 2013.

Kennett, Debbie. *DNA and Social Networking: A Guide to Genealogy in the Twenty-First Century.* The Stroud, UK: History Press, 2011.

Kennett, Debbie. *The Surnames Handbook, a Guide to Family Name Research in the 21st Century.* UK: The History Press, 2012.Klitzman, Robert L. *Am I My Genes: Confronting Fate and Family Secrets in the Age of Genetic Testing.* New York, NY: Oxford University Press, 2012.

Manco, Jean. *Ancestral Journeys: The Peopling of Europe from the first Venturers to the Vikings.* New York, NY: Thames & Hudson, 2013.

McKissick, Katie. What's in Your Genes: Your Genetic Traits. Avon, MA: Adams Media, 2014.

Michaelis, Ron C., Robert G. Flanders, and& Paula H. Wulff. *A Litigator's Guide to DNA: From the Laboratory to the Courtroom.* Amsterdam, Netherlands: Academic Press, 2008.

Moffat, Alistair, and James F. Wilson. *The Scots: A Genetic Journey.* UK: Birlinn, Ltd., 2012.

Pääbo, Svante. Neanderthal Man: In search of lost genomes. New York: Basic Books, 2014.

Pomery, Chris. *Family History in the Genes: Trace Your DNA and Grow Your Family Tree.* Kew, UK: National Archives. 2007.

Preer, Jean. *Library Ethics.* Westport, CT: Libraries Unlimited, 2008.

Quackenbush, John. *The Human Genome: The Book of Essential Knowledge.* Watertown, MA: Charlesbridge, 2011.

Redmonds, George, Turi King, and David Hey. *Surnames, DNA, and Family History.* New York, NY: Oxford University Press, 2011.

Schonberger, Viktor Mayer, and Kenneth Cukier. *Big Data: A Revolution That Will Transform How We Live, Work, and Think.* New York, NY: Houghton Mifflin Harcourt, 2013.

Skloot, Rebecca. *The Immortal Life of Henrietta Lacks.* New York, NY: Broadway Paperbacks, 2010. (The paperback edition is preferred because of the supplementary material added at the end.)

Smolenyak, Megan, and Ann Turner. *Trace Your Roots with DNA: Using Genetic Tests to Explore Your Family Tree.* Emmaus, PA: Rodale, 2004.

Spinello, Richard A. *CyberEthics: Morality and Law in Cyberspace*, 4th ed. Sudbury, MA: Jones and Bartlett, 2011.

Sykes, Bryan. *DNA USA: A Genetic Portrait of America.* New York, NY: Liveright, 2012.

Sykes, Bryan. *Saxons, Vikings and Celts: The Genetic Roots of Britain and Ireland.* New York, NY: Norton, 2006.

Sykes, Bryan. *The Seven Daughters of Eve.* New York, NY: Norton, 2001.

Watson, James D., with Andrew Berry. *DNA: The Secret of Life.* New York, NY: Knopf, 2003.

Wells, Spencer. *Deep Ancestry: Inside the Genographic Project.* Washington, DC: National Geographic, 2007.

Wells, Spencer. *The Journey of May, a Genetic Odyssey*, New York, Random House, 2003.

Wheelwright, Jeff. *The Wandering Gene and the Indian Princess: Race, Religion, and DNA.* New York, NY: Norton, 2012.

BLOGS

Bettenger, Blaine. *The Genetic Genealogist: Adding DNA to the Genealogist's Toolbox.* http://www.thegeneticgenealogist.com/.

Dowell, David R. *Dr. D Digs up Ancestors.* http://blog.ddowell.com.

Estes, Roberta. *DNAeXplained—Genetic Genealogy: Discovering Your Ancestors—One Gene at a Time.* http://dna-explained.com/.

Kennett, Debbie. *Cruwys News: The Day-to-Day Activities of the Cruwys/Cruse One-Name Study with Occasional Diversions into Other Topics of Interest such as DNA Testing and Personal Genomics.* http://cruwys.blogspot.com.

Moore, CeCe. *Your Genetic Genealogist.* http://www.yourgeneticgenealogist.com.

Pontikos, Dienekes. *Dienekes' Anthropology Blog.* http://dienekes.blogspot.com/

Russell, Judy G. *The Legal Genealogist.* http://www.legalgenealogist.com/blog/. (Judy discusses DNA-related topics in her blog posts each Sunday. Posts on other days are devoted to legal topics.)

ONLINE RESOURCES

GEDMatch.com: http://www.GEDMatch.com.

Hill, Richard. "Guide to DNA Testing: How to Identify Ancestors, Confirm Relationships, and Measure Ethnic Ancestry through DNA Testing." Version 2.O. http://bit.ly/1qsaiEs

International Society of Genetic Genealogy (ISOGG):

- Autosomal DNA testing comparison chart: http://www.isogg.org/wiki/Autosomal_DNA _testing_comparison_chart.
- For DNA-newbies: http://www.isogg.org/newbiemenu.htm.
- "DNA Testing Guide": http://www.dna-testing-adviser.com/DNA-Testing-Guide.html.
- "mtDNA Testing Comparison Chart": http://www.isogg.org/wiki/MtDNA_testing_comparison _chart.
- "Y-DNA Testing Comparison Chart": http://www.isogg.org/wiki/Y-DNA_testing_comparison _chart.
- "Y-DNA Haplogroup Tree 2013": http://www.isogg.org/tree/index.html.

Janzen, Tim, Emily Aulicino, and Lisa McCullough. "Basics of Chromosome Mapping," presentation to GFO DNA Interest Group Meeting, July 27, 2013: http://www.gfo.org/ intgrp/chromo-mapping.pdf.PhyloTree.org. "mtDNA tree": http://www.phylotree.org/tree/ main.htm.

U.S. National Library of Medicine (NLM), National Institutes of Health (NIH). "Genetics Home Reference: Your Guide to Understanding Conditions": http://ghr.nlm.nih.gov.

Udacity. "Tales from the Genome": https://www.udacity.com/course/bio110.

University of Utah, Genetic Science Learning Center: http://learn.genetics.utah.edu/content/ chromosomes/.

Wheaton, Kelly. "Beginners Guide to Genetic Genealogy," *Wheaton Surname Resources*: https:// sites.google.com/site/wheatonsurname/beginners-guide-to-genetic-genealogy/.

Glossary

adenine (A)   One of the four chemical bases of DNA. It always pairs with thymine on the opposite strand along the double helix.

admixture   "Genetic admixture occurs when individuals from two or more previously separated populations begin interbreeding." [*Wikipedia*]

allele   Alternative form of a gene. Alleles are found at the same place on a chromosome. Generally, there is a paternal allele and a maternal allele.

autosome   A numbered chromosome. A chromosome not involved in gender determination.

base pair   "Two chemical bases bonded to one another forming a 'rung of the DNA ladder.' " [*Talking Glossary of Genetic Terms* from the National Human Genome Research Institute]

Cambridge Reference Sequence (CRS)   Based on the sequence of the first mitochondria to be mapped. This work was done at Cambridge University and became the reference set against which other mitochondrial results were measured.

celibate DNA   DNA that is passed intact from one of the parents without mixing with DNA from the other parent. Example: Y-DNA from father to son(s) and mtDNA from mother to children of both genders.

centiMorgan (cM)   A unit of measurement based on recombination. It is used to measure genetic linkage, or how likely two alleles are to be inherited together.

chromosome   Structure within the nucleus of the cell containing DNA. A normal human cell contains 46 chromosomes arranged into 23 pairs.

clade   A group of individuals who descend from a common ancestor.

cM   See *centiMorgan.*

coding region   A sequence of DNA that can result in the production of a protein necessary for cellular function.

CRS   See *Cambridge Reference Sequence.*

cytosine (C)   One of the four chemical bases of DNA. On the double helix, it is paired with guanine (G).

DNA   Deoxyribonucleic acid; an infinitely detailed instruction book on how to create our specific physical body, to change it over time, and if necessary to repair it.

dominant   A phenotype that is expressed regardless of what the opposite allele codes for. In pea plants, the yellow allele is dominant over the green allele, so that all offspring of a homozygous yellow × green cross will be yellow pea plants.

double helix   Two linear strands of DNA that bond together in a manner that resembles a twisted ladder.

founder effect   A mutation found in high frequency in a specific population due to the contribution of a single individual.

GEDCOM   A file format for sharing genealogical information from one program to another.

gene   The basic physical unit of inheritance passed from parent to child; it contains instructions for the production of proteins.

genealogical time   The time when most of us have had surnames and some of us have a chance to find documentary records mentioning our ancestors. For those of European descent, that period is generally from the 17th century to the present.

genetic distance   FTDNA uses genetic distance in its yDNA and mtDNA test results to say how many markers are different between two people. 23andMe uses the term to describe the number of centiMorgans two people share.

genetic drift   Random fluctuation in genetic values from generation to generation due to chance events.

genome   An entire set of instructions found in a single cell.

genome-wide association studies (GWAS)   Studies that compare large groups across their genomes to identify differences as well as commonalities that may be linked to observed differences (e.g., medical conditions).

genotype   Observable traits of an individual that are attributable to genes.

guanine (G)   One of the four chemical bases of DNA. On the double helix, it is paired with cytosine (C).

haplogroup   An ancient grouping to which a male or a female ancestor belonged. Each individual has a maternal and a paternal haplogroup. The classification schemes for maternal and paternal haplogroups are completely separate.

haplotype   A group of alleles of different genes on a single chromosome that are frequently inherited usually as a unit.

Human Genome Project (HGP)   A massive public and private effort to create the first comprehensive draft of a map of the human genome.

hypervariable region (HVR)   A region on the mitochondria that seems to mutate more rapidly than other regions.

HVR1   There are two mitochondrial hypervariable regions used in human mitochondrial genealogical DNA testing. HVR1 is considered a "low-resolution" region. Getting HVR1 and HVR2 DNA tests can help determine one's haplogroup. HVR1 locations are numbered 16001–16568. [*Wikipedia*]

HVR2   There are two mitochondrial hypervariable regions used in human mitochondrial genealogical DNA testing. HVR2 is considered a "high-resolution" region. Getting HVR2 DNA tests can help determine one's haplogroup. HVR2 locations are numbered 001–574. [*Wikipedia*]

IBD   See *identical by descent*.

IBS   See *identical by state*.

identical by descent (IBD)   Identical DNA segments shared because individuals descend from a *demonstrable* common ancestor.

identical by state (IBS)   Segments that are a result of a common ancestor, but because that common ancestor lived so far back in time, we cannot *demonstrate* who that person was. Some segments appear to be IBS due to incorrect phasing of alleles; these segments are a result of the technology used in DNA testing, and are neither IBD nor IBS.

insertion   A letter of the genetic code that is added to the genome. A special kind of mutation.

locus   Specific location on a chromosome.

marker   A DNA segment with a known location.

mitochondrial DNA (mtDNA)   Free-floating circular rings of DNA contained inside the mitochondria of the cell. Each cell may have hundreds or thousands of copies. Each person inherits mtDNA from his or her mother but men do not pass it on.

Mitochondrial Eve   The most ancient woman who had at least two daughters who have current living direct female lines. Not to be confused with Biblical Eve.

modal haplotype   The most commonly occurring value for given locations along the genome.

most recent common ancestor (MRCA)   The most recent ancestor whom two individuals share in common.

MRCA   See *most recent common ancestor.*

Mutation   A change in value at a location on the genome

noncoding DNA   DNA not known to be tied to any trait or function.

non-paternity event (NPE)   A situation in which the surname of the biological parent is not the name passed on the offspring. It can result from unrecorded adoptions, legal name changes, births out of wedlock, and so on.

NPE   See *non-paternity event*.

nuclear DNA   DNA within the nucleus of cells. The 22 pairs of autosomal chromosomes plus the two sex chromosomes.

Nucleus   A membrane within a cell that contains the chromosomes and most of the genetic material.

personalized medicine   Medical treatment that is individualized based on the patient's genomic information.

Phenotype   Observable traits of an individual.

population bottleneck   A situation in which the choice of mates is greatly restricted because of geographic isolation, famines, wars, religious beliefs, or other factors. This results in pairing with mates who are related.

promiscuous DNA   DNA that becomes diluted in each intergenerational transfer because a contribution of each parent is included and only half of the parential DNA is passed on. See also *recombining*.

rCRS   See *Revised Cambridge Reference Sequence.*

Recessive   A phenotype that is expressed only if both alleles are identical.

Recombining   Mixing process of paternal and maternal DNA within the nuclear chromosomes except the sex chromosomes.

Reconstructed Sapiens Reference Sequence (RSRS)   A system that uses a reconstructed Mitochondrial Eve as the starting point for describing how mitochondria differ.

Revised Cambridge Reference Sequence (rCRS)   A system that uses a slightly revised version (1998) of the first mitochondria sequences identified at Cambridge University as the point for describing how mitochondria differs.

RSRS   See *Reconstructed Sapiens Reference Sequence.*

sex chromosome   The 23rd pair of chromosomes in the nucleus, which may be either two X chromosomes or a combination of an X chromosome and a Y chromosome. This pattern determines the gender of the individual.

short tandem repeat (STR)   A repeating pattern of a genetic code of letters at a location on the genome. The value reported is the number of times that pattern is repeated. Pronounced "stir."

single nucleotide polymorphism (SNP)   A single change in the DNA bases at a given location. Pronounced "snip."

SNP   See *single nucleotide polymorphism.*

sperm   Contains the father's DNA contribution to offspring.

STR   See *short tandem repeat.*

surname   Family name.

thymine (T)   One of the four chemical bases of DNA that pairs with adenine (A) bases on the opposite strand along the double helix.

TMRCA   Time to the most recent common ancestor.

umbilical DNA   Mitochondrial DNA that passes down through maternal lines.

X chromosome   A "sex chromosome" within in the cell nucleus. Two copies, one from each parent, are present in females. Males have only one copy, which is contributed by the mother.

Y chromosome   The shortest of the 46 chromosomes within the nucleus of a cell of males.

Y-Chromosome Adam   The most ancient man who had two or more sons, each of whom has an unbroken line of male descendants down to the present. Not to be confused with Biblical Adam.

A more complete glossary, "Genetics Home Reference: Your Guide to Understanding Genetic Conditions," can be found at http://ghr.nlm.nih.gov/glossary.

See also Sorenson Molecular Genealogy Foundation, "Glossary," http://www.smgf.org/glossary.jspx.

Index

protocols of males, 94, 102; approximate percentage inherited by female from each ancestor, 96; approximate percentage inherited by male from each ancestor, 97; four simple inheritance rules, 98; of male first cousins, 100–101; research techniques, 103; sons' inability to inherit through father, 96, 97

yAdam, 108, 116
Y chromosome, 8; (illustration), 7. *See also* yDNA

Y-Chromosome Adam. *See* yAdam
yDNA, 17–43; characteristics, 13; descendant tree, 30; inheritance, 60; surname projects, 28–30, 112; future exploration, 155. *See also* Celibate DNA; guY chromosome; guY DNA; Y chromosome
Years per generation, 85
Ysearch.org, 22–23, 37
ySNPs, 39–40; terminal, 155

Zero–sum game, 131, 138–39

About the Author

DAVID R. DOWELL, PhD, is an author and genealogist who has been researching his family history since the 1960s. He is the manager of two surname and one haplogroup DNA projects. In addition he was an academic librarian for 35 years at Iowa State University, Duke University, Illinois Institute of Technology, Pasadena City College, and Cuesta College. The former chair of the Genealogy Committee of the American Library Association, Dowell belongs to the Southern California Genealogical Society, the New England Historic Genealogical Society, the Middle Tennessee Genealogical Society, the National Genealogical Society, and the International Society of Genetic Genealogists. He holds multiple degrees in history and in library science and served as a special investigative officer in the Air Force. His blog, *Dr. D Digs up Ancestors*, can be found at http://blog.ddowell.com. His previous publications include *Crash Course in Genealogy* (Libraries Unlimited, 2011).